Δ The Triangle Papers: 61

Energy Security and Climate Change

A Report to
The Trilateral Commission

North American and Lead Author
JOHN DEUTCH

European Author
ANNE LAUVERGEON

Pacific Asian Author
WIDHYAWAN PRAWIRAATMADJA

Published by
The Trilateral Commission
Washington, Paris, Tokyo
2007

The Trilateral Commission was formed in 1973 by private citizens of Europe, Japan, and North America to foster closer cooperation among these three democratic industrialized regions on common problems. It seeks to improve public understanding of such problems, to support proposals for handling them jointly, and to nurture habits and practices of working together. The Trilateral countries are nations in Europe, North America, and Pacific Asia that are both democratic and have market economies. They include the member and candidate member nations of the European Union, the three nations of North America, Japan, South Korea, the Philippines, Malaysia, Indonesia, Singapore, Thailand, Australia, and New Zealand.

These essays were prepared for the Trilateral Commission and are distributed under its auspices. They were discussed at the Commission's annual meeting in Brussels on March 18, 2007.

The authors—from North America, Europe, and Pacific Asia—have been free to present their own views. The opinions expressed are put forward in a personal capacity and do not purport to represent those of the Trilateral Commission or of any organization with which the authors are or were associated.

Library of Congress Cataloging-in-Publication Data

Deutch, John M., 1938-
Energy security and climate change / John Deutch, Anne Lauvergeon, and Widhyawan Prawiraatmadja.
p. cm.
ISBN 978-0-930503-90-1
1. Energy policy--Environmental aspects. 2. Energy development--Environmental aspects. 3. Climatic changes. I. Lauvergeon, Anne. II. Prawiraatmadja, Widhyawan. III. Trilateral Commission. IV. Title.
HD9502.A2D456 2007
363.738'74561--dc22

2007019437

The Trilateral Commission

www.trilateral.org

1156 15th Street, NW
Washington, DC 20005

5, Rue de Téhéran
75008 Paris, France

Japan Center for International Exchange
4-9-17 Minami-Azabu
Minato-ku
Tokyo 106, Japan

Contents

Preface v

The Authors vii

1 Priority Energy Security Issues 1
John Deutch

2 Energy Security and Climate Change: A European View 51
Anne Lauvergeon

3 Pacific Asia Energy Security Issues 81
Widhyawan Prawiraatmadja

Preface

The three papers on energy security and climate change contained in this report were delivered to the Trilateral Commission by their authors on March 18, 2007. Taken together, they lay out the physical constraints and political circumstances that govern national choices for improving energy security and address the daunting challenge of mitigating global climate change.

What is perhaps not surprising about the three presentations—one by a former leading U.S. policymaker, the second by the leader of a major international nuclear company, and the third by the head of business planning for a large national oil company—is that they agree on the key facts and circumstances affecting the world's energy future. What is most striking to the reader is that they disagree so little in describing the limited policy choices available to Trilateral countries for managing extremely complex and interconnected energy problems facing both developed and emerging economies.

Although the structure of the report is three individual papers encompassing both global and regional views, there is surprising congruence. The first, by principal author John Deutch, former U.S. director of central intelligence and under secretary of energy, offers a tour d'horizon of global energy issues and climate change that also provides analysis of U.S. domestic energy politics. The second, by Anne Lauvergeon, chief executive officer of Areva, the French nuclear company, offers the European viewpoint on these issues and makes a detailed and forceful case for ambitious nuclear power development within an enhanced nonproliferation framework. The last, by Widhyawan Prawiraatmadja, the head of corporate planning and business development for the Indonesian national oil company, PT Pertamina, proposes mechanisms for increasing regional energy security that take into consideration Pacific Asia's broad geography, fast-paced development, and infrastructure requirements.

The singular contribution of the Trilateral Commission and of the high-level task forces it has created over the years to address thorny megaproblems is once again demonstrated in this report. What is most useful about its approach is both a concentration on careful analysis of global remedies and an appreciation of the different regional approaches that can be devised to meet common challenges.

Thus, we see in the United States a slow pace in adopting sustained policy measures that address the challenge of limiting greenhouse gas emissions and embarking on the long process of making a transition from a petroleum-based economy. In Europe, different political forces have combined to produce greater commitment to energy pricing mechanisms and greenhouse gas mitigation, but also great interest and appreciation of the need for concerted international cooperation to address widespread solutions, that is, those involving participation by the United States and rapidly developing economies. In Pacific Asia, the supply concerns of rapidly expanding economies such as China's demonstrate the difficulties of consistent market protections as well as the problems the Trilateral countries face in finding solutions that will effectively address climate change without disrupting economic activity in both the developed and developing worlds.

All three authors see the concept of energy security as a process of managing rather than eliminating or even drastically reducing the risk inherent in dependence on imported energy supplies. What is most hopeful, given their clear appreciation and analysis of regional pressures that affect each of the Trilateral regions, is their unambiguous appreciation that both energy security and some successful mitigation of greenhouse gas emission growth can be achieved, provided they are addressed with a global outlook and effective international solutions. Each nation-state continues to reserve to itself key decisions affecting itself, but no one who has read this report can doubt that a comprehensive, cooperative, and worldwide approach to the latter is the only hopeful path that can lead to needed control of climate-changing emissions stemming from human activity.

This report is being published in advance on the eve of the 2007 Group of Eight summit in Germany as a contribution to the debate on energy security and climate change and with the hope that it will find a place high on the summit agenda.

The Authors

North American and Lead Author

John Deutch is an Institute Professor at the Massachusetts Institute of Technology (MIT). Previously he served as director of central intelligence (1995–96), deputy secretary of defense (1994–95), and under secretary of defense for acquisition and technology (1993–94). Dr. Deutch has also served as director of energy research (1977–79), acting assistant secretary for energy technology (1979), and under secretary (1979–80) in the U.S. Department of Energy. He has served on the President's Nuclear Safety Oversight Committee (1980–81), the President's Commission on Strategic Forces (1983), the White House Science Council (1985–89), the President's Intelligence Advisory Board (1990–93), the President's Commission on Aviation Safety and Security (1996), the President's Commission on Reducing and Protecting Government Secrecy (1996), the President's Council of Advisers on Science and Technology (1996–2000), and as chairman of the Commission to Assess the Organization of the Federal Government to Combat Proliferation of Weapons of Mass Destruction (1998–99). Dr. Deutch has been a member of the MIT faculty since 1970 and has served as chairman of the Department of Chemistry, dean of science and provost. He has published more than140 technical publications in physical chemistry as well as numerous publications on technology, international security, and public policy issues. Dr. Deutch holds a B.A. in history and economics from Amherst College, a B.S. in chemical engineering from MIT, and a Ph.D. in physical chemistry from MIT. He is a member of the board of directors of Citigroup, Chinere Energy, Cummins Engine, Raytheon, and Schlumberger.

European Author

Anne Lauvergeon has served as chief executive officer of Areva group since 2001 and chairman and chief executive officer of Areva NC group since 1999. She holds a degree in physics and is a graduate of the French National School of Mining Engineering (Ecole des Mines) and the French Ecole Normale Supérieure. She started her professional career in 1983 in the iron and steel industry and moved afterward to Usinor.

In 1984, she directed the European safety studies for the chemical industry of CEA (Commissariat à l'Energie Atomique, the public technological research organization in France). From 1985 to 1988, she supervised the underground utilities activities in and around Paris and was appointed deputy director of the General Mining Council in 1988. In 1990, she was named adviser for economic international affairs at the French Presidency and deputy chief of its staff in 1991. At the same time she became "sherpa" to the president in charge of the Group of Seven summit preparations. In 1995, she became a partner of Lazard Frères & Cie in Paris, spending several months in their New York offices. In 1997, she joined Alcatel Telecom as senior executive vice president and was appointed a member of the Executive Committee in July 1998. She was in charge of international organizations and the group's interests overseas in the energy and nuclear fields. She currently serves as an administrator for Suez, Total, Safran, and Vodafone.

Pacific Asian Author

Widhyawan Prawiraatmadja joined PT Pertamina (Persero), the Indonesian state oil company, in January 2005 and currently serves as head of corporate planning and business development. He is also chairman of the executive board of the Foundation of Indonesian Institute for Energy Economics (IIEE). Prior to joining Pertamina, he was director and senior associate of Fesharaki Associates Consulting and Technical Services Inc. (FACTS Inc.) and visiting fellow at the East-West Center (EWC). FACTS Inc. is a prominent consulting firm specializing in downstream oil and gas in the east-of-Suez region (the Middle East and the Pacific Asia region) whereas EWC is a research institution that studies linkages between the United States and the Pacific Asia region. Dr. Prawiraatmadja spent more than fifteen years with both institutions in Hawaii before deciding to return to Indonesia. Prior to his living abroad, he was with PT Redecon, a consulting firm based in Jakarta, in which he served as energy division manager. Dr. Prawiraatmadja specializes in energy economics, notably in downstream oil and gas in the Pacific Asia region and the Middle East. He has conducted research and consulting projects on the commercial energy sector relating to economic, environmental, and national policy issues. His area of expertise includes national energy-economic policy, petroleum and natural gas market analysis, petroleum refining economics, interfuel substitution, environmental issues related to fossil fuel use, and energy modeling. He has several publications and journal articles and has been cited in the press and industrial media. He has done research and consulting work on global and regional issues as well as specific countries such as Australia, Bahrain, Brunei, China, India, Indonesia, Iran, Japan, Korea, Kuwait, Malaysia, Philippines, Qatar, Saudi Arabia, Singapore, Taiwan, Thailand, United Arab Emirates, and the United States. He holds Ph.D. and M.A. degrees in economics from the University of Hawaii and an industrial engineering degree from Bandung Institute of Technology.

1

Priority Energy Security Issues

John Deutch

Energy markets create economic interdependence among Trilateral countries and between Trilateral countries and the rest of the world. Energy is an important domestic political issue because our economies rely on access to dependable supplies of energy and because consumers and economies are sensitive to energy costs. Economies can prosper when energy costs move higher, but the reality and perception of price instability create uncertainty that affects consumer spending and dampens investment. Thus, domestic energy policies have international consequences, and international developments affect domestic economies.

The term "energy security" is intended to convey the connection between the economic activity that occurs in both domestic and international energy markets and the foreign policy response of nations (apart from the fundamental connection between national security and a healthy economy). Increasingly, both governments and the public recognize that the linkage to national security matters must be evaluated alongside economic considerations in adopting energy policies. For example, efforts to prevent Iran's nuclear program from leading to a nuclear weapons capability, taken together with the importance that Iranian oil exports (now about 3 million barrels per day) have for the world oil price, and the potential for Iran to heighten or dampen civil violence and unrest in Iraq and elsewhere in the Middle East vividly illustrate the difficulty and complexity of the energy security linkages.

The energy issue is not new to the Commission. In 1998 the Trilateral Commission published a comprehensive energy report authored by William F. Martin, Ryukichi Imai, and Helga Steeg, entitled *Main-*

Author's note: I am grateful to Harold Brown, Henry Jacoby, Paul Joskow, and Arnold Kanter for helpful comments on earlier drafts of this manuscript.

taining Energy Security in a Global Context,[1] and at the 2006 Tokyo Plenary Meeting, Steve Koonin spoke about available technology choices for meeting future energy needs.[2] In 2007, in Brussels, the Trilateral Commission continues its consideration of energy. This background paper draws on thirty years of involvement with these issues, including as a government official in the U.S. Department of Energy and Department of Defense, research and teaching about energy technology at the Massachusetts Institute of Technology (MIT), and involvement with many private energy firms. This paper aspires to deepen the analysis of some of the key energy security issues we face today:

- Oil and gas import dependence;
- Energy infrastructure vulnerability;
- Global warming; and
- The future of nuclear power.

In addressing each of these four topics, the connection between energy and security, actions that Trilateral countries should take, and the interactions between the four issues are identified.

Before beginning, two points require emphasis: first, progress on each of these issues requires a heightened level of international cooperation; and second, enlightened common action by nations can substantially lower the cost of adapting to our energy future. This is true for Trilateral countries and the international community. Moreover, Trilateral members, in their relationships with their colleagues and their governments, can make a difference in how well and quickly we act.

The United States and, I suspect, most Trilateral countries have made little progress in adopting measures recognized as necessary to address effectively the four key energy security challenges listed above. For example, the United States does not have in place a policy process that harmonizes the foreign and domestic aspects of energy policy. There are two underlying causes. First, progress on each of these key issues requires sustained policies over a long period of time—decades rather

---

1 William F. Martin, Ryukichi Imai, and Helga Steeg, *Maintaining Energy Security in a Global Context* (Washington, D.C.: Trilateral Commission, 1998), www.trilateral.org/projwork/tfrsums/tfr48.htm.

2 Steve E. Koonin, "In Search of New Global Frameworks for Energy Security," in *Challenges to Trilateral Cooperation* (Tokyo: Trilateral Commission, 2006), 3, www.trilateral.org/annmtgs/trialog/trlglist.htm.

than years. As prices and events change, the public's attention and the attention of their elected representatives waxes and wanes. The public memory of Indian, Pakistani, and North Korean nuclear tests dims, while the potentially adverse consequences of each of these nations possessing a nuclear capability do not. Irreversible global climate change will not be apparent until many years after current elected officials leave office, which reduces the incentive to allocate scarce resources for needed investment in mitigating greenhouse gas emissions.

The second and related reason is that elected officials tend to avoid speaking plainly about energy issues. The public understandably wants cheap and dependable energy that permits an improved lifestyle and neither harms the environment nor depends on foreign sources. Simultaneously satisfying all these conditions is difficult, if not impossible, especially since, in a market-based energy economy, energy imports rise when imports are cheaper for the consumer than domestic energy alternatives. To quote my MIT colleague, economist Lester Thurow:

> It is only when we demand a solution with no cost that there are no solutions.

In the United States and, I surmise, elsewhere, political figures seem unable to resist the temptation to tell the public what they want to hear. One hears the call for energy independence—an unattainable concept—and arbitrary goals for renewable energy or efficiency improvements that are not based on realistic assessment of either economics or technology or on a willingness to put in place policy measures such as energy consumption or carbon emission taxes that would catalyze the transformation to a new global system of energy supply and use.[3] It is up to leaders in Trilateral countries to urge their governments to take urgently needed action.

---

3 A particularly embarrassing example for me is one of the new initiatives in "Six for '06" announced by congressional Democrats (http://democrats.senate.gov/agenda/) immediately after their November 2006 midterm election sweep, which states:

> ENERGY INDEPENDENCE—LOWER GAS PRICES: Free America from dependence on foreign oil and create a cleaner environment with initiatives for energy-efficient technologies and domestic alternatives such as biofuels. End tax giveaways to Big Oil companies and enact tough laws to stop price gouging.

Oil and Gas Import Dependence

Import dependence has both economic and political consequences.[4] Here we are concerned with the political consequences that result from both the reality and perception of anticipated economic consequences.

The trend in world oil supply and demand under business-as-usual assumptions is clear.

Demand and Supply of Oil

The U.S. Department of Energy's Energy Information Administration (EIA) projects in the *International Energy Outlook 2006*[5] an increase in world oil consumption from 80 million barrels of oil per day (MMBOD) in 2003 to 118 MMBOD in 2030, that is, an average annual increase of 1.4 percent, accompanied by an uncertain real price increase. (The EIA considers a range of prices from $38 per barrel to $96 per barrel, with $57 per barrel in the reference case; all prices are in real 2004 dollars.)

---

4 This section relies heavily on the recent Council on Foreign Relations report, *National Security Consequences of U.S. Oil Dependency* (New York: Council on Foreign Relations, October 2006), www.cfr.org/publication/11683/. James Schlesinger and I cochaired the independent task force that prepared this report.

5 *International Energy Outlook 2006* (Washington, D.C.: U.S. Department of Energy, Energy Information Administration, June 2006), Chap. 3, http://www.eia.doe.gov/oiaf/ieo/pdf/oil.pdf. Table N1 provides additional information.

Table N1. World Oil Consumption by Region and Country Group, 2003 and 2030, million barrels per day

Regions and country groups	2003	2030
North America	24.2	33.4
Non-OECD Asia	13.5	29.8
OECD Europe	15.5	16.3
OECD Asia	8.8	10.1
Central and South America	5.3	8.5
Middle East	5.3	7.8

Source: *International Energy Outlook 2006* (Washington, D.C.: U.S. Department of Energy, Energy Information Administration, June 2006), DOE/EIA-0484(2006), www.eia.doe.gov/oiaf/ieo/excel/figure_27data.xls.

Asian countries, including China and India, that are not members of the Organization for Economic Cooperation and Development (OECD) account for 43 percent of the increase in consumption.[6] Importantly, EIA projects a non-OECD Asia oil consumption growth rate of 3 percent, so that by 2030, non-OECD Asia will account for about 28.1 percent of world consumption, compared with 18.6 percent in 2003.[7]

Most of the world's oil reserves are in the Middle East and in Organization of the Petroleum Exporting Countries (OPEC), as shown in table 1 on page 6.

Accordingly, importing nations for the foreseeable future will rely in large measure on oil from these countries.

Between 2003 and 2030, the world oil trade is expected to increase:

- In 2003, total world oil trade consisted of 53 MMBOD. Of this amount, 32 MMBOD came from OPEC, including 22.5 from the Persian Gulf region. North America imported 13.5 MMBOD, and non-OECD Asia imported 9.9 MMBOD, with China accounting for 2.8 MMBOD of that total.
- In 2030, it is estimated that total world oil trade will be 77 MMBOD. Of this amount, it is estimated that OPEC will produce 48.5 MMBOD, including 34 MMBOD from the Persian Gulf Region. North America is projected to import 19 MMBOD, and non-OECD Asia 22 MMBOD, with China accounting for 11 MMBOD of that total.

These data suggest why there is increasing concern about the security aspects of dependence on oil and gas imports.

On the demand side, in the absence of an extended global recession, there appears to be no diminution in the pace of increase in world oil consumption. The new, rapidly growing emerging economies such as China and India are becoming major importers of oil. The sharp increase in oil prices that occurred in early 2006 was the first price shock that can be characterized as demand driven; Hurricane Katrina and supply concerns with Nigeria and Venezuela were also factors. The economic consequence is the effect of price shocks on the economies of importing countries, although OECD economies have recently gone through a major price increase with little effect on their economies.

---

6 Ibid., 25.

7 Ibid., 27.

Table 1. World Oil Reserves, by Country, as of January 1, 2006, billion barrels

Country	Oil reserves
Saudi Arabia	264.3
Canada	178.9
Iran	132.5
Iraq	115.0
Kuwait	101.5
United Arab Emirates	97.8
Venezuela	79.7
Russia	60.0
Libya	39.1
Nigeria	35.9
United States	21.4
China	18.3
Qatar	15.2
Mexico	12.9
Algeria	11.4
Brazil	11.2
Kazakhstan	9.0
Norway	7.7
Azerbaijan	7.0
India	5.8
Rest of world	68.1
World total	1,292.5

Source: "Worldwide Look at Reserves and Production," *Oil & Gas Journal* 103, no. 47 (December 19, 2005): 24–25.

On the supply side, importing nations will remain dependent to a large extent on oil coming from politically unstable parts of the world—the Persian Gulf, for example—and from suppliers such as Iran, Russia, and Venezuela that may actively oppose the interests and policies of Trilateral countries. Non-OPEC production between 2003 and 2030 is estimated to fall slightly as a proportion of all exports. The concern here is that effective control of supply and price by a cartel of exporting countries—OPEC—could potentially be used as a political instrument to influence, for example, the Palestine-Israel question. The oil trade transfers significant wealth to producer countries such as Iran that do not share the values or interests of Trilateral countries, and petrodollars can be used to support terrorist organization or efforts to acquire weapons of mass destruction, as the was the case in the 1980s with Libya and Iraq.

In addition, concerns are increasing about the functioning of oil and gas markets, especially because there has been a movement away from transparent markets governed by commercial considerations to state-to-state agreements between the national oil companies (NOCs) of the major resource holders (MRHs) and the new rapidly growing emerging economies.

There has been a major shift in oil reserves and production from the international oil companies (IOCs) to the NOCs. In the early 1970s, the IOCs controlled about 80 percent of reserves and production, while NOCs controlled 20 percent. Today that proportion is about reversed. A 2005 article in the *Washington Post* included a stark graphic that showed the largest non-state-controlled IOC, ExxonMobil, was number fourteen on a list of the top twenty-five MRHs.[8]

While there is a wide variability in the capacity and efficiency of the NOCs to explore, produce, and market their hydrocarbon reserves, it is likely that NOCs will become progressively more important on the supply side of the market. If IOCs are to prosper, they will need to adapt their traditional approach that seeks ownership and control of reserves in MRH countries.

The MRHs are quite clear that they intend to use their resources to advance political objectives. The rhetoric of Iran and Venezuela is especially strident. But Russia has also made plain that centralizing control over its petroleum industry is intended to give Russia political leverage—a message that especially threatens Europe, with its great dependence on Russian gas imports.

The net result of the combination of more muscular NOCs and new consumers that are unsure about the source of their future supply is an increase in state-to-state agreements, with new users seeking to

---

8 Justin Blum, "National Oil Firms Take Bigger Role: Governments Hold Most of World's Reserves," *Washington Post*, August 3, 2005, Sec. D, http://www.washingtonpost.com/wp-dyn/content/article/2005/08/02/AR2005080201978_pf.html.

Figure 1. Trends in Rising Chinese Oil Imports, Prices, and Number of Political Oil Deals

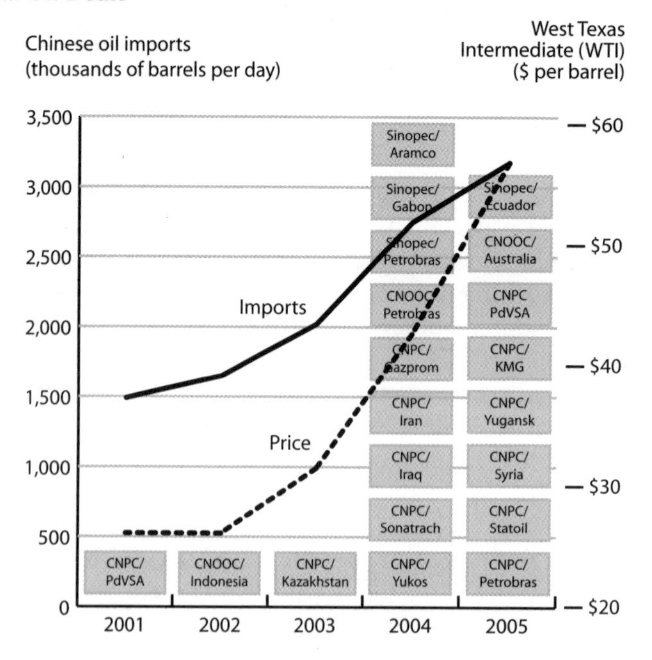

Sources: Sinopec Corp., CNOOC Limited, and China National Petroleum Corporation. Price data from *BP StatisticalReview of World Energy 2006.* Import data from *EIA China Country Analysis Brief,* August 2006. As cited in *National Security Consequences of Oil Dependency* (New York: Council on Foreign Relations, October 2006).

lock up supply. These agreements frequently involve political concessions and nonmarket considerations that are quite different from what is expected in a conventional commercial transaction; the Chinese arrangements in Africa with Sudan and Angola are frequently cited. Figure 1 vividly illustrates the growth in Chinese offshore oil activity.

These cases are the consequence of China's policy of "going out" for resources globally.[9]

---

9 Aaron L. Friedberg, "'Going Out': China's Pursuit of Natural Resources and Implications for the PRC's Grand Strategy," *NBR Analysis* 17, no. 3 (September 2006), 21<em>30, www.nbr.org/publications/issue.aspx?ID=392.

The objection to state-to-state agreements is not that oil is taken off the open market—to date the quantities tied up are small—or that new consumers are paying too high a price to lock up oil supplies, but instead the objection is to the use of oil as a political instrument by those whose political purposes may run counter to the interests of Trilateral countries. For example, Angola provides China with 15 percent of its total oil consumption. In May 2006, Angola's Sonangol and China's Sinopec signed a multibillion-dollar agreement to develop jointly offshore blocs with reserves estimated at 4.5 billion barrels (China beating out India in this bid). Since 2004, in parallel, the Chinese government has extended extensive technical assistance to Angola, including a soft loan of $4 billion and pledges to invest $400 million in Angola's telecommunications sector and to upgrade Angola's military communications network. China imports about 10 percent of its oil from Sudan, where it has major investments; China is reported to be Sudan's biggest supplier of arms and military equipment.

Natural Gas

The outlook for global natural gas demand and supply lags oil in terms of the security concerns based on import dependency, and it shows greater regional variation. In brief, four countries—Russia, Iran, Qatar, and Saudi Arabia—account for 60 percent of world gas reserves. Both OECD and non-OECD countries, especially non-OECD Asia, are projected to increase their consumption of gas over time, increasingly through international trade.

This projected international trade may occur by pipeline, as from Canada to the United States or from Russia to Europe; by liquefied natural gas (LNG), as from Indonesia to Japan or Trinidad to the United States; or by conversion of gas-to-liquefied (GTL), for example, natural gas converted to methanol, in locations such as offshore West Africa, where large reserves of gas are "stranded" far from markets.

Natural gas is an attractive fuel because its production and use is relatively environmentally "clean." The price of natural gas is likely to equilibrate, on average over time, to the price of oil at the point of use—"the burner tip"—because natural gas is a direct substitute for refined oil in industry.

In East Asia, intense competition is likely among Japan, South Korea, Taiwan, and China for available natural gas supplies. China and Japan will compete for control of natural gas pipeline routes from Central Asia to the Pacific to lock in and increase sources of supply.

Europe already is heavily dependent on natural gas imports, especially from Russia. Gazprom has shown its willingness to cut off gas supplies to Ukraine and Belarus on the grounds that the countries are not paying market prices, but the lesson is not lost on Europeans who depend on a reliable supply from Russia. Some of the pipelines that carry Russian gas to Europe transit Ukraine and Belarus, so a dispute between Russia and these countries could easily affect gas delivery to Europe.

North America is certain to become a net importer of natural gas in the near future. The good news is that the natural gas pipeline system and market that serves Canada, Mexico, and the United States has become more integrated. The bad news is that over time North America will increasingly depend on LNG imports. These LNG imports, the source of supply at the margin, will determine (allowing for transportation and processing costs) the price of natural gas in North American markets, as opposed to the cost of North American production.

Effect of Oil and Gas Dependence on International Security

The chronic (and growing) dependence on imported hydrocarbons has many implications for the conduct of foreign affairs by individual nations and for international security.

Because increased demand is recognized as inevitable, at least in the short run, countries will become increasingly intent on assuring a reliable supply and hence sensitive to indications that world oil and gas markets are becoming less open and transparent. Importing countries inevitably will adjust their policies and international relationships to accommodate the interests of those countries that supply their oil and gas. The competition for supply among OECD countries and between OECD and non-OECD countries will increase, giving rise to heightened tensions. Africa and Central Asia will become particular areas for competition. In Central Asia, competition for hydrocarbons and pipeline routes (going east or west) will present Russia and Iran with opportunities to forge new advantageous relationships with China, Japan, India, and others.

Because China is growing so rapidly, its need for hydrocarbon imports will be correspondingly great. Its quest for these resources is sure to add strain in the relations between China and its East Asian neighbors and between China and the United States.[10] The intensely adverse U.S. reaction to the offer by the Chinese National Offshore Oil

Company (CNOOC) to buy the offshore assets of Union Oil of California (UNOCAL) and the incorrect belief that Chinese demand caused the 2005–2006 increase in world prices (or, if you believe this, the more recent decline in prices) indicate how the energy issue can exacerbate an already complicated relationship between these two countries. The U.S. reaction to the CNOOC offer to buy UNOCAL is particularly unfortunate because it contradicts U.S. policy elsewhere in the world of support for opening the oil sector of other countries—for example, Russia—to investment. The truth is that China's approach to its participation in the world oil and gas market is evolving; influencing its evolution is important to Trilateral countries.

Responding to the foreign policy challenges caused by these features of world oil markets would be easy if energy security were the sole or priority concern. But energy security is just one of many foreign policy objectives of Trilateral countries. Our energy security objectives must be balanced against combating terrorism; slowing the spread of weapons of mass destruction; and encouraging democracy and human rights, economic growth, and environmental protection. Energy dependence constrains Trilateral countries in pursuing other important foreign policy objectives.

Response of Trilateral Countries

Any response must be based on three realities.

First, the world is running out of low-cost oil; over time the real price of oil will go up. From time to time the price of oil may decline, but over the long haul, the world is on a staircase of rising prices for hydrocarbon fuel.

Second, Trilateral countries and other large oil-importing countries, such as China and India, will, for at least the next several decades, remain dependent on oil from the Persian Gulf—Iran, Iraq, Saudi Arabia, and Kuwait.

Third, we must begin a transition away from a petroleum economy. This is a long-term problem with no short cuts. Investments must be made today if we are to have choices in the future.

---

10 An important analysis is Kenneth Lieberthal and Mikkal Herberg, "China's Search for Energy Security, Implications for U.S. Policy," *NBR Analysis* 17, no. 1 (April 2006), www.nbr.org/publications/issue.aspx?ID=217.

Foreign policy measures. These three realities point the way for what Trilateral countries should do. I suggest four measures intended to influence international energy developments. Even if successful, taken together these measures serve only to improve our capacity to manage oil and gas import dependence; they do not offer the prospect of eliminating energy dependence or even reducing the expected dependence to a level that qualitatively would change security concerns for the foreseeable future.

1. Trilateral countries have common interests with the new, large, emerging economies. This means the International Energy Agency (IEA) should be broadened to include new significant consumers such as China and India because, ultimately, all consumers will benefit from a level playing field where there is competition for resources on commercial terms.
2. When expanded, the IEA should address common policies with regard to national stockpiles and response to price shocks. IEA members should continue to advocate that countries not subsidize internal oil and gas prices. Permitting prices to rise to world levels is a necessary, but perhaps not a sufficient, step toward limiting demand growth. If there are groups within a country—for example, low-income families and the elderly—that are especially hurt by higher energy prices, individual countries will, and should, adopt targeted assistance programs rather than further distort markets.
3. Trilateral countries have an interest in maintaining and increasing oil and gas production everywhere in the world.

    a. Trilateral countries should work together to encourage stability in the Persian Gulf. This means that diplomacy, trade, and economic policies need to balance the important objective of continued production with other objectives such as human rights and democratization.

    b. Trilateral countries should continue to encourage production in non-OPEC countries. This has long been an objective of OECD countries and has met with limited success. The proportion of oil produced by non-OPEC countries is unlikely to increase dramatically, but the effort should continue.

    c. Trilateral countries need to encourage production where possible in their own countries. For example, Canada's huge tar sands resources (330 billion barrels) are expected to reach a production level of between 2 and 4 million barrels per day in the next fifteen years.[11] Production should also be encouraged in the North Sea. The United States should also increase domestic oil and gas production from some areas in Alaska, the Gulf of Mexico, and the Atlantic and Pacific coasts that are currently off-limits because of environmental concerns. While incremental U.S. production will be only a small part of total supply, it is difficult to see how the United States or other Trilateral countries can convince others to expand production without making any effort to increase production at home.

4. Trilateral countries should encourage responsible governance in producing countries in West Africa generally and in Ecuador in South America. The motivation here is not altruism but rather that political and social stability are necessary for continued, even expanded, oil and gas production. Stability requires some use of oil revenues to improve the economic and social circumstances of ordinary people. The expanding energy sector in Africa presents significant challenges.[12]

The leverage of Trilateral countries on international energy developments is limited. In part the limitation follows from dependence and in part from the fact that energy is only one of many foreign policy objectives. Some advance the notion that Trilateral countries can and should adopt more aggressive policies, such as by establishing a linkage between cooperative behavior on both energy and nonenergy matters by a producer country, and access to technology, domestic markets, and trade with the importing countries. There may be particular situations where such a tactic might work to advantage, but the approach is unlikely to be widely effective and it would be unwise because it is a move away from open and transparent world markets.

---

11 See "CAPP Releases 2006 Canadian Crude Oil Forecast," Canadian Association of Petroleum Producers, May 17, 2006, www.capp.ca/default.asp?V_DOC_ID=1169. Current tar sands production by strip mining and in situ methods such as steam-assisted gravity drive (SAGD) is about 900,000 barrels per day.

12 See *More Than Humanitarianism: A Strategic U.S. Approach toward Africa* (New York: Council on Foreign Relations, January 2006), http://www.cfr.org/publication/9302/more_than_humanitarianism.html. The report includes a description of Chinese activities in Africa.

Domestic policy measures. While Trilateral country leverage on international oil matters may be limited, Trilateral countries can do a lot more with domestic policies. Trilateral countries should be focused on adopting domestic policies that begin the long process of moving away from a petroleum-based economy. I suggest three priority domestic policy measures.

1. The highest priority should be to maintain a high price on liquid fuel, because this encourages efficiency and fuel switching, dampens demand, and stimulates innovation. High liquid fuel prices are in place in Europe and Japan, but not in the United States.[13] I favor adoption of an additional tax in the range of $1.00 per gallon imposed on motor gasoline, diesel, and other petroleum

13 Figure N1 from Cambridge Energy Research Associates vividly makes the point that the United States (and China) lag behind the rest of the world in petroleum taxes.

Figure N1. Gasoline Prices and Taxes in Selected Countries, 2006

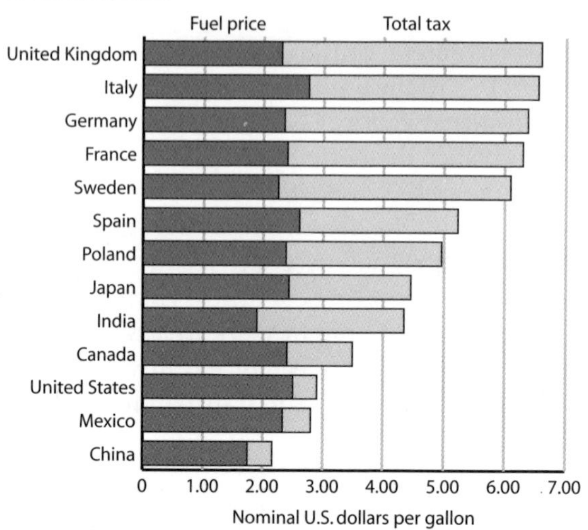

Sources: Cambridge Energy Research Associates; *Energy Prices and Taxes,* International Energy Agency, third quarter 2006. See http://www2.cera.com/ gasoline/press/.

Notes: Japan and China prices are for 91 RON unleaded. India price is for 91 RON leaded. Canada price is for 92 RON unleaded. U.S. price is for 87 octane (R+M)/2 basis. Data are third quarter 2006 averages, as available.

products at a time when pump prices are falling, so the impact on the public will be less. A tax of this level would raise considerable revenue, in excess of $150 billion per year, which should be allocated for three purposes: countervailing reduction in other taxes; increased support for energy research, development, and demonstration (RD&D); and impact assistance for those most adversely affected by the tax.

Many will note the political difficulty, if not impossibility, of the U.S. Congress assessing such a tax; thus, there is interest in alternative approaches such as tradeable gasoline rights[14] or tightening of present corporate average fuel economy (CAFE) standards. CAFE standards, because they mandate fuel economy, only indirectly reduce gasoline consumption. Some combination, rather than any one of these three measures, may be more politically feasible.

2. The second priority of Trilateral countries should be to adopt a much larger and more ambitious RD&D effort to create future options for new liquid fuels or substitutes for liquid fuels. One approach is to develop new technologies that use these fuels more efficiently. The other approach is to develop new technologies for alternatives to liquid fuels. Three deserve mention: synthetic liquids and gas from shale and coal; biofuels such as ethanol from biomass; and alternative nonfossil, electricity generation-based transportation systems.

   a. *Synthetic liquids and gas from shale and coal.* As conventional, low-cost sources of oil and gas are depleted, there will be a steady progression to more costly fossil sources of liquid fuels. The first stage will be unconventional oil and gas resources, such as coal bed methane and tar sands. The next stage will use the considerable shale and coal resource base to produce synthetic fuels. I was deeply involved in the launch of the ill-fated U.S. Synthetic Fuels Corporation of the 1970s, and today's efforts can learn much from this experience. Figure 2 gives a highly schematic view of how synthetic fuels are produced.

---

14 My friend and distinguished Trilateral Commission member, Martin Feldstein, is the leading proponent of this approach; see "Tradeable Gasoline Rights," *Wall Street Journal,* June 5, 2006, www.nber.org/feldstein/wsj060506.html.

Figure 2. System Elements for Production of Synthetic Fuels from Coal, Natural Gas, and Biomass

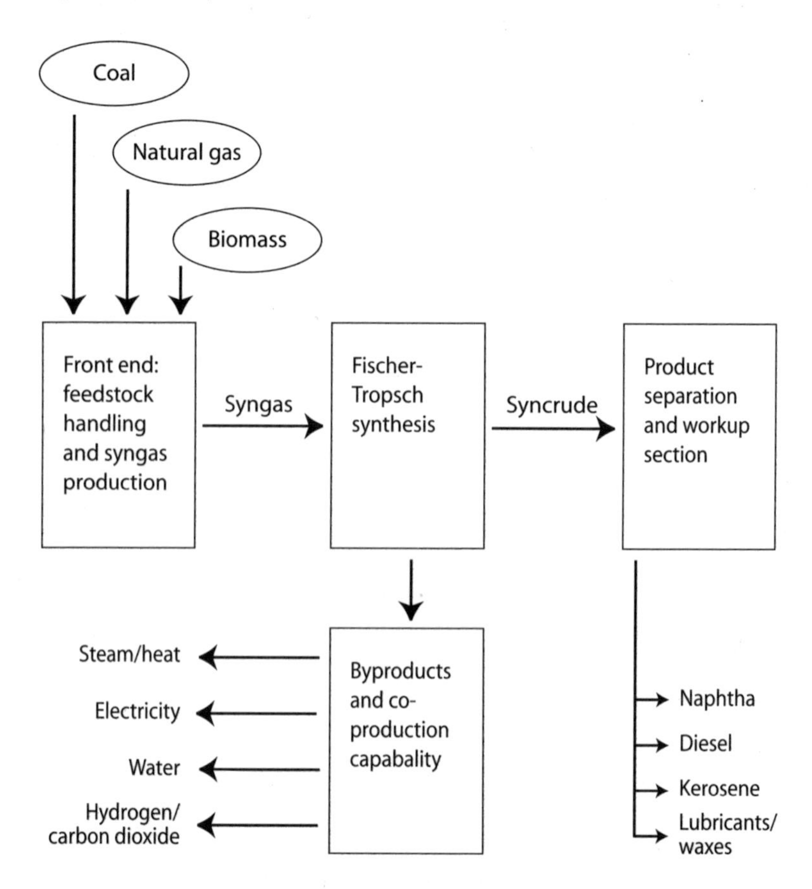

Source: *Annual Energy Outlook 2006, with Projections to 2030*, report no. DOE/EIA-0383(2006) (Washington, D.C.: U.S. Department of Energy, Energy Information Administration, February 2006), 54, figure 19, http://www.eia.doe.gov/oiaf/aeo/pdf/0383(2006).pdf.

Synthetic fuels face two challenges. The first is cost. The capital cost is high, in the range of \$50,000–\$75,000 per barrel per day capacity, which in turn leads to high product costs. For example, a first-of-a-kind shale plant has been estimated to be able to produce synthetic liquid in the range of \$70–\$95 per barrel (2005 dollars) over the life of the plant.[15] The cost of initial plants to produce synthetic liquids from coal will be in a comparable range, depending on coal cost and quality. As industry capacity expands and there is learning by doing, these costs should come down, perhaps by $20–$30 per barrel, as industrial capacity expands.

The second challenge to synthetic-fuels production from shale and coal is environmental. These conversion projects will require attention to air and water quality, waste material disposal, and land remediation. On a large scale, carbon dioxide (CO_2) emissions are also of concern. The conversion of coal to synthetic oil, for example, involves the formation of between two and three molecules of CO_2 for every atom of carbon in the oil.[16] Thus, the CO_2 emissions of synthetic oil can be double or more (after by-product credit) compared with conventional oil. If (as discussed later) global constraints on carbon emissions are adopted in order to reduce the threat of global warming, carbon capture and sequestration (CCS) might be required when producing synthetic fuels from coal and shale, driving costs much higher.

The *Annual Energy Outlook 2006* (published by the Energy Information Administration of the U.S. Department of Energy), in the high price case, assuming the use of underground mining with surface retorting, estimates that U.S. oil shale production will begin in 2019 and grow to 410,000 barrels per day by 2030.[17] *Annual Energy Outlook 2006* projects U.S. coal-to-liquids production in the range 800,000 to 1.7 million barrels per day in 2030, depending upon oil price assumptions. Worldwide coal-to-liquids production in 2030 is estimated to be in the range of 1.8 to 2.3 million barrels per day.[18] If shale oil production includes CO_2 capture, the cost rises substantially.

15 James T. Bartis et al., *Oil Shale Development in the United States: Prospects and Policy Issues* (Santa Monica, Calif.: RAND Corp., 2005), www.rand.org/pubs/monographs/2005/RAND_MG414.pdf.

16 I stress that CO2 emissions from synthetic-fuels production depend on the technology employed.

17 *International Energy Outlook 2006*, 54.

18 Ibid., 55.

b. *Biofuels.* Biofuels from biomass also have significant potential to displace a portion of petroleum-based liquid fuels. In countries that have a highly industrialized agricultural sector, the production of ethanol or biodiesel from food crops will not be economic without government subsidies. Moreover, although it remains hotly debated in the United States, ethanol produced from corn or sugar likely requires two-thirds of a barrel of the oil equivalent of the natural gas and oil needed to produce one barrel of oil equivalent ethanol (after allowance for by-product credits) because of the energy intensity of cultivation and energy requirement for fermentation and distillation.

In countries with a more favorable climate and a less energy-intensive agricultural sector, such as Brazil, the economics of conversion of food crops to biofuels may be different than in the United States. The United States, rather foolishly, places 5.5 cents per gallon tariff on both sugar and ethanol imports in order to protect U.S. ethanol distillers and corn farmers from this competition.

The situation with regard to the potential for the production of biofuels such as ethanol or butanol from cellulosic biomass, such as agricultural waste, corn stover, switch grass, and poplar, is quite different. These crops are fast growing and are not cultivated in an energy-intensive way, neither do they command the high price of a food crop. Thus, there is the potential for economic production of biofuels. The biomass can be converted to liquid fuel in two ways. The first is indirectly through gasification, as indicated in figure 2.

The second approach uses modern biotechnology to engineer new organisms that will efficiently and economically digest the cellulose and hemicellulose into usable liquid products. (Native organisms easily digest the starch-based sugars in food-based crops.) This approach is receiving great attention today, but there are technical challenges. For fermentation, cellulosic materials require severe conditions to separate the cellulose and hemicellulose from the feed-starting material. The biotechnology and metabolic engineering required to produce biofuels remain to be demonstrated on an industrial scale. Several corporations, including BP, Chevron, and DuPont, have large programs, and in the United States many biotech startups are exploring various aspects of this biomass-to-

Figure 3. Oil Alternatives: Costs and Emissions Vary Widely

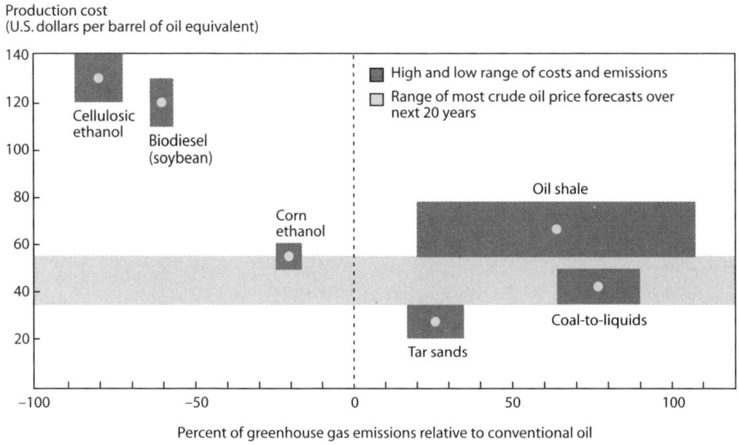

Source: Richard G. Newell, "What's the Big Deal about Oil: How We Can Get Oil Policy Right," *Resources,* No. 163 (Fall 2006): 9, www.rff.org/Documents/Rff-Resources-163.pdf.

biofuels approach. Under optimistic assumptions, the cost per barrel oil equivalent for cellulosic ethanol in the future is in the range of $40 per barrel, so there is genuine reason for enthusiasm here.[19]

Annual Energy Outlook 2006 projects 700,000–900,000 oil equivalent barrels per day of U.S. ethanol production and 1.7–3.0 million barrels per day oil equivalent (including biodiesel) worldwide production in 2030, depending upon world oil prices.[20] There are limits, however, to ultimate production—perhaps 30 million barrels per day worldwide—because of land and water availability. Of course, aquaculture is another potential source of biomass.

A recent publication by Resources for the Future provides a useful summary of the range of estimates of the costs and greenhouse gas emissions of liquid fuel alternatives relative to conventional oil (figure 3).

---

19 See John Deutch, "Biomass Movement," *Wall Street Journal,* May 10, 2006, http://online.wsj.com/article/SB114722621580248526.html.

20 *International Energy Outlook 2006,* 58.

c. *Alternative electricity-based transportation systems.* Alternative electricity-based transportation systems offer another path to replacing liquid-fueled transportation systems. Both mass transit rail-based systems and electric hybrid or all-electric cars are interesting possibilities; the latter would benefit greatly from an improvement in batteries or other methods of electricity storage.

This pathway, of course, trades off petroleum dependence for electricity generation. I discuss later the security concerns from coal-fired electricity generation (global warming) and from nuclear power (proliferation).

3. The third domestic priority for Trilateral countries is to explore new ways of managing the energy RD&D process. Successful innovation in the energy sector requires a significant research and development effort, accompanied by a demonstration stage undertaken for the purpose of demonstrating the technical feasibility, cost, and environmental character of new technology. The demonstration phase is necessary because in most OECD countries, energy production and distribution are done by the private sector. Private firms and the financial institutions that provide firms with the capital needed for the massive investments required will not adopt unproven technology. Some form of government assistance is likely to be necessary for first-of-a-kind plants.[21]

The mechanism for public support for technology change of the kind that is needed differs among Trilateral countries. The European Union, Japan, and the United States have very different procedures for deciding how to share the costs of RD&D between the government and the private sector. Nevertheless, there may be attractive opportunities for cooperation among Trilateral countries—one long-term example is cooperation on fusion energy research.

---

21 Much has been written about how the process of government encouragement of civilian technology might be improved. An old but nevertheless still relevant discussion is given in *The Government Role in Civilian Technology: Building a New Alliance,* the report of a panel chaired by Harold Brown (Washington, D.C.: National Academies Press, 1992), http://books.nap.edu/catalog.php?record_id=1998.

Energy Infrastructure Protection

As energy use expands and resources originate at progressively greater distance from users, the energy infrastructure that supports energy distribution becomes more vulnerable to damage from nature, technical failure, or human causes.

Natural disasters. As low-cost oil and natural gas resources are depleted, production facilities move to more extreme environments such as production platforms operating in the Arctic or offshore in deep water. Transportation facilities, collection systems, and pipelines must follow the production platforms. These facilities are vulnerable to extreme natural phenomena such as hurricanes and earthquakes, as Hurricanes Katrina and Rita demonstrated in the Gulf of Mexico in 2005.

Technical failure. Technical failure is a term that refers to interruptions or accidents arising from human or natural causes in the operation of an element of the energy infrastructure. As this infrastructure becomes larger, more complex, and dispersed, such events are inevitable. There are many recent examples: oil spills from pipelines and tankers, transmission grid failures, and accidents in refineries. Unquestionably, safety and reliable operation will receive greater attention by both industry and regulators. Efforts to improve safety and reliable operation for normal commercial operation will benefit efforts to protect the energy infrastructure from natural disasters and hostile threats.

Terrorist, insurgency, and hostile-state threats to the energy infrastructure are likely to grow.[22] Because much of the energy infrastructure is located in remote areas or in areas such as the Middle East that are politically unstable, it is reasonable to expect an increased number of attacks. In February 2006, for example, terrorists made an abortive attack on the 600,000 barrel per day Abqaiq oil processing center in Saudi Arabia. In September 2006, terrorists believed to have connections with Al Qaeda simultaneously attacked a refinery and an oil storage depot in Yemen.

---

22 The distinction between terrorist and counterinsurgency threats is blurred, but there are many examples: Chechnya, Colombia, Sudan, Angola, Nigeria, and Iraq are prominent among them.

It is not only oil and gas facilities that are vulnerable, but also tankers, port facilities, offshore production platforms, pipelines, power plants (especially nuclear power stations), and electricity transformation and transmission networks. And what about the ships that transport nuclear fuel and separated plutonium around the world? The reason that the energy infrastructure is an attractive target to terrorists is that these targets are "soft," that is, easily destroyed or incapacitated by a cyber attack that penetrates the SCADA (Supervisory Control and Data Acquisition) computer systems that do real-time monitoring and control of plant and equipment. The destruction of such targets can cause tremendous disruption and economic loss without large loss of life—a characteristic that can be very attractive to organized terrorist groups that seek to achieve political objectives and wish to avoid acts that invite more extreme retaliation.

These vulnerabilities—natural, technical, and from terrorists and other groups—give rise to security concerns that are receiving greater attention from both industry and governments.

Civilian responses of Trilateral countries. Trilateral countries are likely to pay considerably greater attention in the future than in the past to the vulnerability of the energy infrastructure, and they will adopt measures that better protect key facilities and plant operations from both natural disaster and terrorist attack. While it is not possible to guarantee absolute security from an attack, it is possible to take steps that will make this infrastructure more secure and raise the cost of a successful attack. Such protection is expensive, however, and arriving at a reasonable level will require cooperation between industry and government. Energy firms, especially those with international operations, should expect to spend more time on emergency preparedness planning: evaluating the vulnerability of their facilities and operations to natural disasters and terrorist attack and making investments in systems and procedures for protection.

Effective warning and defense systems will require international cooperation. For example, consider that LNG requires a liquefaction facility, an LNG tanker, and a re-gasification facility that spans two countries and open ocean transport. This points to what Trilateral countries should do:

- Establish international standards for the siting, construction, and operation of facilities;

- Exchange best practices information on energy infrastructure operations;
- Undertake joint operations to improve infrastructure protection, especially customs and port security; and
- Practice and exercise defenses and recovery.

Role of military forces. It is worth noting that deployed military forces help protect energy infrastructure. Military cooperation often offers a practical means of technical information exchange and joint planning and exercises in, for example, port security, air traffic control, and telecommunications. In general, cooperation between the military forces of Trilateral countries and the military forces of MRH countries, when it occurs, encourages professionalism and hence more responsible conduct by local military. There are additional, more central, connections between military force deployment and economic security.

The most obvious example is the role the U.S. Navy plays in keeping sea lanes safe for international shipping. Most nations recognize and welcome the function that the U.S. Navy plays in maintaining open seas. However, China and perhaps other nations will worry about the capability of the U.S. Navy to block tankers and other shipping entering or leaving Chinese ports, which may encourage China to begin the lengthy, expensive, and potentially risky process of developing a bluewater navy capability.

Most fundamentally, deployed military forces, if used wisely, can contribute to regional political stability. As the experience of the U.S. military intervention in Iraq indicates, military deployment does not automatically lead to stability; intervention can bring unexpected and costly consequences. Nevertheless, Trilateral countries, facing many decades of dependence on imported oil and gas, should consider how deployed military forces and their operations should be used in a manner that contributes to the objective of maintaining stable supply. For example, some will argue that the U.S. military should maintain a significant force deployment in both the Middle East and East Asia because this presence contributes to regional stability and thus will be generally welcomed by governments in the region. Forward-deployed military forces advance the U.S. interests of maintaining stability in oil-producing regions and countering terrorism and proliferation.

Global Warming

Global warming is a different kind of foreign policy issue. It does not have the direct national security implication, for example, of war in the Persian Gulf. But global warming is arguably, along with global poverty, the issue that can most seriously affect the economic and social circumstances of future generations.

Although not all agree, the informed scientific consensus is that the consequences of global warming are likely to be very damaging if anthropogenic emissions of greenhouse gases continue on their present course and are not reduced.[23] I have followed the evolution of understanding about the implications of greenhouse gas emission for climate change since I was director of energy research in the U.S. Department of Energy in the 1970s. I believe that continued emission of greenhouse gases will cause an increase in global temperature, although the timing and amount of the increase is somewhat uncertain. The impact of the temperature increase on climate and the ability of economies and societies to respond (there will be winners and losers) is less sure. Global warming will occur. We should adopt policies now to reduce emissions—how stringent depends upon judgments about present and future costs. The longer the world waits to adopt carbon constraints, the more difficult and costly it will be for our economies to adapt.

It is mindless to deny the foreign policy implications of a situation where business-as-usual conduct by individual nations involves the common welfare of all. Moreover, the global warming issue divides Trilateral nations, especially the United States and Europe, as to what should be done. Global warming also divides OECD countries and the rapidly growing, large emerging economies over who should bear the cost of mitigation. This subject is sure to remain prominently on the international agenda in years ahead. If the United States or any other OECD country that is a large producer of greenhouse gas emissions is to retain a leadership role in other areas, it cannot just opt out of the global climate change policy process.

---

23 The leading international authority on global warming is the Intergovernmental Panel on Climate Change (IPCC). Much useful information is found on its Web site, www.ipcc.ch/.

Table 2. CO_2 Emissions by Region

Year	OECD	Non-OECD	Total
2003	3.59	3.4	6.83
2030	4.77	7.14	11.91

Source: *International Energy Outlook 2006* (Washington, D.C.: U.S. Department of Energy, Energy Information Administration, June 2006).

Outlook for Global CO_2 Emissions

There are many greenhouse gases,[24] but I will focus on carbon dioxide, CO_2, because this product of combustion from fossil fuels, especially coal, accounts for over 70 percent of all greenhouse gas emissions, of which about 40 percent is from coal combustion, primarily from electricity generation.

The anticipated growth in these CO_2 emissions is given in table 2.

During the period 2003–2030, the *International Energy Outlook 2006* reference case projects that the CO_2 emissions of OECD countries will grow by 1.1 percent per year, while non-OECD Asia will grow by 3.6 percent.[25]

Because of the considerable lag between emissions and atmospheric concentration response, even if the world reduced emissions today, it would be a long time before atmospheric concentrations stabilized. The Intergovernmental Panel on Climate Change (IPCC), under the auspices of the World Meteorological Organization (WMO) and the UN Environmental Program (UNEP), offers a striking illustration (figure 4) of this lag in the results of a model that compares an emissions trajectory that stabilizes CO_2 atmospheric concentrations at 550 parts per million (ppm), about twice the preindustrial levels; this concentration would result in a global average increase of about 2.2°C. This trajectory, although uncertain, should be compared with the model prediction of continual upward trend in temperature, if the world stabilized emissions at the year 2000 level.

---

24 The principal greenhouse gases are: CO_2, carbon dioxide; CH_4, methane; N_2O, nitrous oxide; PFCs, perfluorocarbons; HFCs, hydrofluorocarbons; SF6, sulphur hexafluoride. Each compound has a different global warming potential.

25 *International Energy Outlook 2006*, 73, table 12.

Figure 4. Impact of Stabilizing Emissions versus Stabilizing Concentrations of CO_2

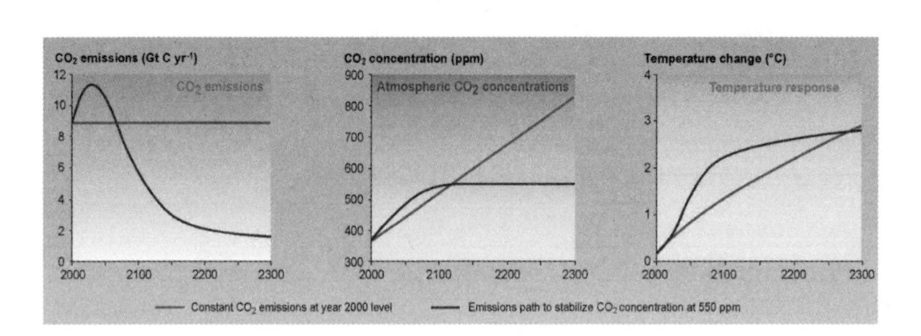

Source: *Climate Change 2001: Synthesis Report* (Geneva: Intergovernmental Panel on Climate Change, 2001).

Please remember that the relationship projected between the atmospheric concentration and global mean average temperature increase is based on a model that cannot be completely validated empirically. Thus, today researchers are addressing a more sophisticated question: What is the probability that the temperature increase will be greater or less than the 2.2°C predicted in the mode?

What would it take to reduce carbon emissions? At MIT, we have just completed a study, *The Future of Coal: Options for a Carbon Constrained World,*[26] that used the MIT Emissions Prediction and Policy Analysis (EPPA) model[27] to analyze the level of carbon emission reduction needed to stabilize world emissions by 2050. This is only a step toward the goal of stabilizing CO_2 atmospheric concentrations at 550 ppm. While emissions are sharply reduced compared with business-as-usual, further reductions would be required. The MIT EPPA model is a self-consistent description of economic adjustments that

26 S. Ansolabehere et al., *The Future of Coal: Options for a Carbon Constrained World* (Cambridge: Massachusetts Institute of Technology, 2007), http://web.mit.edu/coal/.

27 A description of the MIT Emissions Prediction and Policy Analysis (EPPA) model is found in Sergey Paltsev et al., *The MIT Emissions and Policy Analysis (EPPA) Model: Version 4,* Report no. 125 (Cambridge, Mass.: Joint Program on the Science and Policy of Climate Change, August 2005), http://web.mit.edu/globalchange/www/MITJPSPGC_Rpt125.pdf.

Figure 5. Scenarios of Penalties on CO_2 Emissions, dollars per ton CO_2 in constant dollars

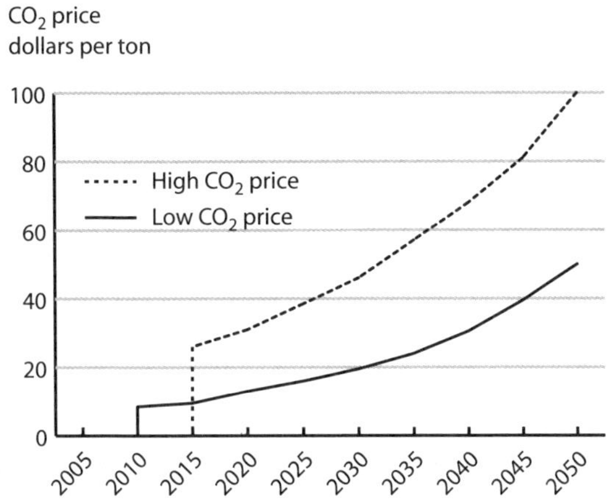

Source: S. Ansolabehere et al., *The Future of Coal: Options for a Carbon Constrained World* (Cambridge: Massachusetts Institute of Technology, 2007), 9, Fig 2.2, http://web.mit.edu/coal/.

occur over time by region and industrial sector, based on assumed policies, supply and demand curves for commodities, and technical characteristics of energy technologies.

For the MIT *Future of Coal* study, the EPPA model was used to estimate the future effects of two carbon emission price penalty scenarios. This penalty or emissions price can be thought of as the result of a global cap-and-trade regime, a system of harmonized carbon taxes, or even a combination of price and regulatory measures that combine to impose marginal penalties on emissions. The result is presented in figure 5 for assumed real price penalties placed on CO_2 emissions.

If such a pattern of CO_2 emission penalties were adopted, global CO_2 emissions would be stabilized by mid-century (see figure 6).

The low CO_2 price case resembles the recommendation of the recent National Commission on Energy Policy;[28] the effect of this low-price scenario lags the high-price scenario by about twenty-five years.

---

28 *Ending the Energy Stalemate: A Bipartisan Strategy to Meet America's Energy Challenges* (Washington, D.C.: National Commission on Energy Policy, December 2004), www.energycommission.org/files/contentFiles/report_noninteractive_44566feaabc5d.pdf.

Figure 6. Global CO_2 Emissions under Alternative Policies with Universal, Simultaneous Participation, Limited Nuclear Expansion, and EPPA-Ref Gas Prices, $GtCO_2$/year

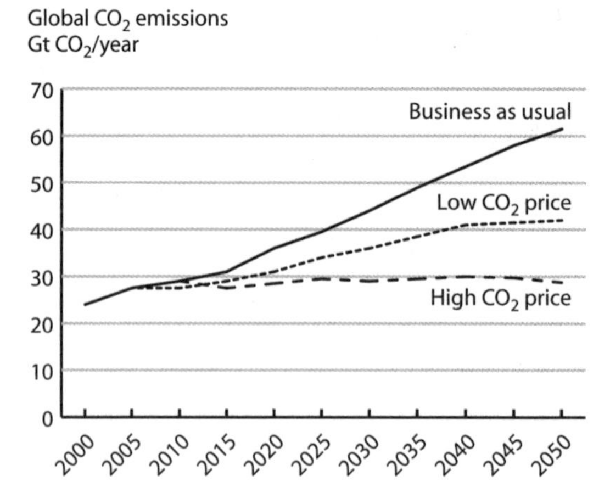

Source: S. Ansolabehere et al., *The Future of Coal: Options for a Carbon Constrained World* (Cambridge: Massachusetts Institute of Technology, 2007), 10, Fig 2.3, http://web.mit.edu/coal/.

This analysis shows that it is possible to stabilize global CO_2 emissions by mid-century. Emission reductions will occur because the global economy will respond to the higher price of carbon emissions in three ways: significant reduction in energy use through improved efficiency of energy use and lower demand; a switch to lower carbon-emitting alternatives; and adoption of new carbon-avoiding technologies. For example, in the EPPA model projections, nuclear power, to the extent it is available, will displace coal-fired electricity generation. The United States and the rest of the world will produce significant quantities of biofuels from biomass, about 20 million barrels of oil per day equivalent. Although not modeled, presumably if international carbon credits are traded, there will be an incentive to increase biomass production globally.

The adjustment of global primary energy consumption to higher carbon prices displayed as reductions from a reference case with no prices is given in figure 7 for the case of expanded worldwide nuclear deployment.

Figure 7. Global Primary Energy Consumption under High CO_2 Prices (expanded nuclear generation and EPPA-ref gas prices)

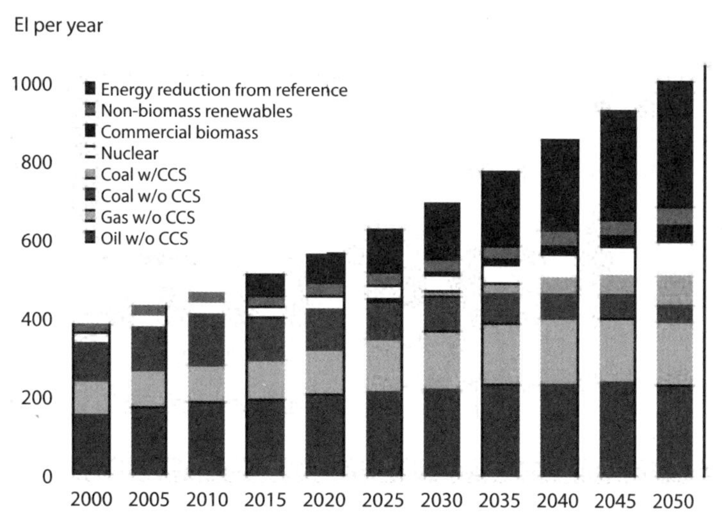

Source: S. Ansolabehere et al., *The Future of Coal: Options for a Carbon Constrained World* (Cambridge: Massachusetts Institute of Technology, 2007), 11, Fig 2.5, http://web.mit.edu/coal/.

Effect on Coal

Coal costs about $1 per million BTU compared with natural gas at about $8 per million BTU, and there are vast deposits of coal in large energy-consuming countries, notably Australia, China, India, Russia, and the United States. Each year, commitments are made that inevitably result in additional future annual emissions of CO_2. For example, China is building more than one large coal (1000 MWe) plant per week, each of which emits approximately 30,000 metric tons of CO_2 daily during the plant's forty-year life. As the use of coal for electricity generation expands significantly, the question arises, what is the future of coal if carbon constraints are applied compared with a business-as-usual world without constraints?

The MIT study, *Future of Coal,* estimates that at a carbon emission price of about $30 (in 2005 dollars) per ton of CO_2, coal combustion to produce electricity with CCS is economic. A snapshot at mid-century shows the positive impact on increased coal use and reduced CO_2 emissions from CCS if the technology is available when a carbon price is

Table 3. Exajoules of Coal Use (EJ) and Global CO_2 Emissions (Gt/yr), 2000 and 2050, with and without carbon capture and storage

	Present course		Limited nuclear		Expanded nuclear	
Coal use	2000	2050	With CCS	Without CCS	With CCS	Without CCS
Global	100	448	161	116	121	78
United States	24	58	40	28	25	13
China	27	88	39	24	31	17
CO_2 emissions: global	24	62	28	32	26	29
CO_2 emissions from coal	9	32	5	9	3	6

Source: S. Ansolabehere et al., *The Future of Coal: Options for a Carbon Constrained World* (Cambridge: Massachusetts Institute of Technology, 2007), xi, Table 1, http://web.mit.edu/coal/.

Notes: Assumes universal, simultaneous participation, high CO_2 prices, and EPPA-ref gas prices. CCS = carbon capture and storage.

imposed (table 3). In 2050, the availability of CCS means that coal use increases more than 80 percent if a high carbon price is imposed, and total CO_2 emissions are reduced more than 10 percent. Under this assumed carbon emission price scenario, moreover, the carbon capture penetration increases rapidly after 2050.

Thus, demonstrating the feasibility of CCS is important for establishing a technical option for CO_2 emission reduction in the future should serious carbon emission control measures be adopted. Today, the leading technologies for coal combustion with CO_2 capture are the integrated coal gasification combined cycle, favored in the United States, and the oxygen-fired, ultra-supercritical, pulverized-coal combustion, favored in Europe. With a CCS charge, the cost of electricity at the bus bar is increased about 50 percent, resulting in an increase in retail electricity cost of about 25 percent.

Because no coal plants currently operate with carbon capture, it is too early to pick a technology "winner," although many do so; moreover, coal type is an important factor in the technology choice. The production of synthetic liquids and gas from oil and shale discussed in the previous section could also involve CO_2 capture in an emission control regime.

Figure 8. Schematic Diagram of Possible CCS Systems

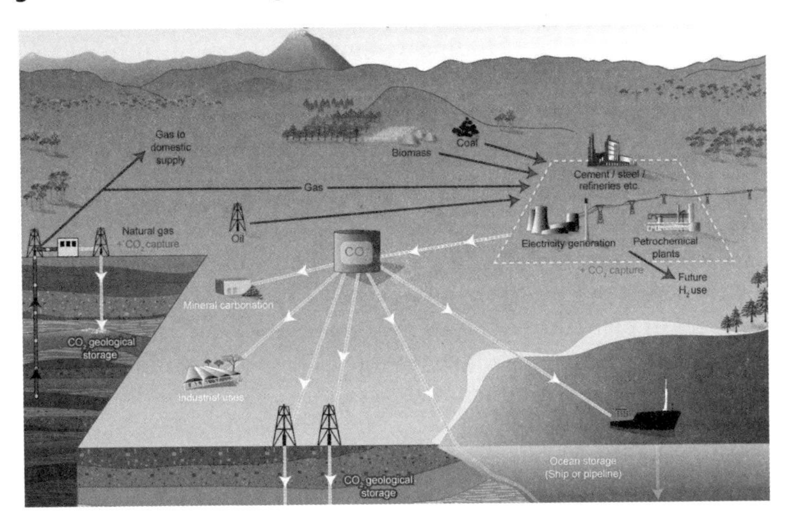

Source: *Climate Change 2001: Synthesis Report* (Geneva: Intergovernmental Panel on Climate Change, 2001).

Status of Sequestration

Technical descriptions of CO_2 sequestration can be found in the IPCC study, *Carbon Dioxide Capture and Storage,*[29] and the MIT study, *Future of Coal.*[30] A CO_2 sequestration system that operates worldwide will have enormous scale—transporting and injecting volumes of CO_2 greatly in excess of the natural gas produced worldwide. Figure 8 indicates the complexity of the process.

The requirements for successfully demonstrating the option of carbon sequestration are three:

1. Integrated operation of capture, transportation, and injection of CO_2 at a storage site;
2. Operation at the scale of at least 1 million tons of CO_2 per year, including a system for measurement, monitoring, and verification; and

---

29 Bert Metz et al., eds., *IPCC Special Report on Carbon Dioxide Capture and Storage* (New York: Cambridge University Press, 2005), www.ipcc.ch/activity/srccs/SRCCS.pdf.

30 Ansolabehere et al., *The Future of Coal.*

3. Establishment of an institutional and regulatory framework that addresses criteria for site selection, injection, monitoring, and operating standards, including assignment of liability provisions for industry and government extending to the end of the life of the storage site; such a framework is essential to establish public acceptance of sequestration, and allowance must be made for differing regulatory practices in different political jurisdictions.

The three major CO_2 sequestration projects[31] currently under way in Sleipner, Norway; Weyburn, Saskatchewan, Canada; and in Salah, Algeria, do not meet these requirements. A number of projects in various stages of planning anticipate integrating CO_2 capture and sequestration. In Germany, Vattenfall is undertaking a program with EU support for operation of an integrated Oxy pulverized coal (lignite) plant with CO_2 capture by 2015.[32] Each of these projects has been designed for a different purpose, and although valuable information has and will be learned, the projects do not satisfy the three requirements needed to establish carbon capture as an acceptable technical, economic, and political option. The annual project cost of each integrated carbon capture and demonstration project should be about $50 million per year.

Five or six integrated sequestration projects should be immediately undertaken to demonstrate that CO_2 sequestration is a credible carbon emission mitigation option. This is a central recommendation of the MIT coal study and it certainly is a program that should be possible for Trilateral countries to accomplish individually and cooperatively. Even the current U.S. administration, which does not believe that carbon emission control is needed, should support projects to establish that the sequestration option is available, if needed, in the future.

How can convergence between developed and developing economies be achieved? The foregoing discussion assumes that there is universal compliance in a carbon control regime. However, the 1994 United Nations Framework Convention on Climate Change[33] and the 1997

31 A brief description of these projects can be found at the MIT Carbon Capture and Sequestration Technologies Web site, http://sequestration.mit.edu/index.html.

32 A description of the Vattenfall plant is found at http://www2.vattenfall.com/www/co2_en/co2_en/index.jsp.

33 Background information and relevant documents can be found at the Web site of the United Nations Framework Convention on Climate Change, http://unfccc.int.

Figure 9. Global CO_2 Emissions under BAU and Alternative Scenarios for Non-Annex B Accession to the High CO_2 Price Path

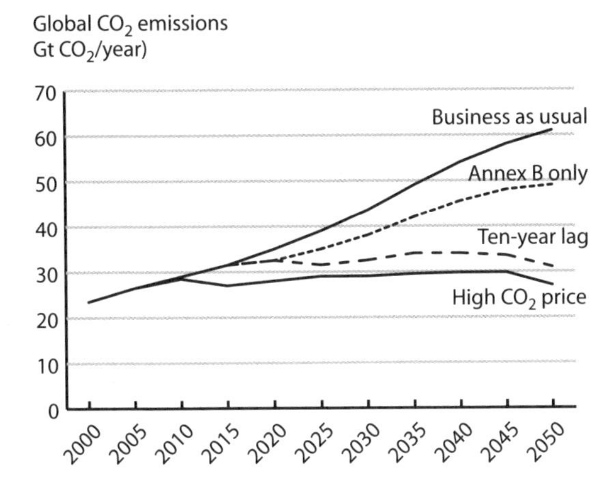

Source: S. Ansolabehere et al., *The Future of Coal: Options for a Carbon Constrained World* (Cambridge: Massachusetts Institute of Technology, 2007), 14, Fig 2.7, http://web.mit.edu/coal/.

Kyoto Protocol include obligations of only thirty-five developed economies (Annex I countries) to limit their emissions to amounts listed in Annex B. Fifteen EU countries agreed to an aggregate reduction of −8 percent of 1990 emissions by 2008/2012; Japan and Canada agreed to −6 percent, with the United States indicating that it would not ratify the protocol and thus would not seek to achieve its previously stated target of −5 percent reduction.

The Kyoto Protocol does not include any obligation on the part of the large, rapidly growing emerging economies to restrict greenhouse gas emissions. This difference in obligations between developed and developing economies reflects a basic difference in interests: developed economies have been responsible for the bulk of past emissions into the atmosphere and wish to constrain future emissions; developing economies, which have not been large emitters in the past and have much lower emissions per capita, argue that in fairness they should have the opportunity of a period of time for economic growth without restrictions on their greenhouse gas emissions. The trouble is that if developing economies do not constrain their emissions, global warming will result regardless of the action taken by the developed economies, as indicated in figure 9 above.

Figure 10. Scenarios of Penalties on CO_2 Emissions: High Price for Annex B Nations and Two Patterns of Participation by Non-Annex B Parties

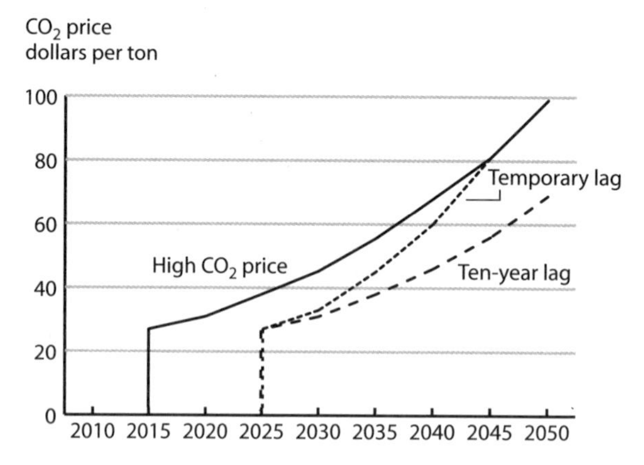

Source: S. Ansolabehere et al., *The Future of Coal: Options for a Carbon Constrained World* (Cambridge: Massachusetts Institute of Technology, 2007), 13, Fig 2.6, http://web.mit.edu/coal/.

The trend is clear: if only Annex B countries constrain emissions and developing countries do not, stabilization of global CO_2 emissions by 2050 is not possible. Of course, it is not necessary for developed and developing economies to adopt exactly the same schedule of restrictions on greenhouse gas and CO_2 emissions. The figure also indicates the consequences of a hypothetical ten-year lag in developing countries accepting a high carbon price of emissions. If developing economies adopt a CO_2 price with ten-year lag, stabilization is possible, depending upon the precise price trajectory. As indicated in figure 10, the lag could be temporary, in which case, during a convergence period, developing economies would experience a higher rate of growth of the real price increase than developed economies. If the lag were permanent, developing economies would have a permanent comparative advantage in energy costs.

How might convergence be achieved? Several possible approaches are discussed as a means of achieving convergence.[34] One possibility is to build on the Kyoto process and pursue continued dialogue in the regularly scheduled Conference of Parties, taking advantage of provisions in the Kyoto Protocol, such as "Clean Development Mechanisms,"

"Joint Implementation," emissions trading, and expanding CO_2 sinks by reforestation. However, we should not expect that continued dialogue based on the Kyoto Protocol will necessarily lead to progress on the underlying equity issue on how global emission constraint cost might be shared between developed and developing economies. Many believe a new and broader framework is needed. At the Trilateral Commission's 2006 North American regional meeting in Cambridge, Massachusetts, Harvard professor Robert Stavins presented a thorough discussion of the architecture needed for a post-Kyoto era.[35] The study, *Beyond Kyoto,* sponsored by the Pew Center on Global Climate Change, is also relevant.[36]

In 2005, Sir Nicolas Stern prepared a review of the economics of climate change for the Chancellor of the Exchequer of the United Kingdom.[37] The Stern review is a comprehensive economic analysis, and it eloquently calls for immediate and collective action. There are several

34 My discussion of convergence addresses the need to harmonize carbon emission constraint policies between developed and emerging economies. For these policies to be effective, if adopted in a developing country, the emerging economy must have a sufficiently developed market economy so that the economic behavioral response assumed for a developed economy—that is, demand response to price changes—operates. If the market structure is not sufficiently developed, as might be the case in China, compliance will not necessarily result in the emission reductions predicted by the conventional models that are calibrated on the response observed in developed economies.

35 Robert N. Stavins, "Climate Change: Technology and Politics" (paper delivered to the Trilateral Commission's 2006 North American regional meeting), www.trilateral.org/NAGp/regmtgs/06pdf_folder/Stavins.pdf. See also Robert N. Stavins and Sheila M. Olmstead, "An International Policy Architecture for the Post-Kyoto Era," *American Economic Review Papers and Proceedings* 96, no. 2 (May 2006): 35–38; Robert N. Stavins, "Forging a More Effective Global Climate Treaty," *Environment Magazine,* December 2004, 24.

36 Joseph E. Aldy et al., *Beyond Kyoto: Advancing the International Effort against Climate Change* (Arlington Va.: Pew Center on Global Climate Change, December 2003), www.pewclimate.org/docUploads/Beyond%20Kyoto%2Epdf.

37 Nicholas Stern, *The Economics of Climate Change: The Stern Review* (Cambridge: Cambridge University Press, 2006), www.hm-treasury.gov.uk/independent_reviews/stern_review_economics_climate_change/stern_review_report.cfm.

technical critiques of the Stern review analysis. For example, William Nordhaus observes that the Stern review's economic analysis adoption of a near zero social discount rate is the crucial determinant of the very high carbon charge the review recommends.[38] I find its discussion of how higher prices will result in the introduction of carbo-avoiding technologies overly optimistic.[39]

Part IV of the Stern review enumerates the advantages and disadvantages of different price-based instruments for influencing carbon emissions. One can imagine how these might be applied to achieve a new international understanding about how convergence should be reached between Annex I (developed) and non-Annex I (emerging economies) on sharing carbon constraint costs. One possible mechanism would be to grant emerging economies initial emission allowances in excess of emission levels in a baseline year.[40] In a global cap-and-trade system, developed countries could purchase emission rights from developing countries, thus effectively providing a transfer of resources to meet the costs associated with carbon emission constraints.

There are several reasons why this approach is unlikely to work. Significant assistance in meeting compliance cost would involve a huge wealth transfer that most developed nations would be reluctant to consider. Emerging economies have given no indication as to what might be an acceptable level of compensation for constraints, and developed countries have given no indication of what consideration might be offered to emerging economies that place constraints on emission profiles. Moreover, it is not clear that all the large emerging economies have the capacity to operate a cap-and-trade system that requires, among other things, a mechanism for internal allocation of emission allowances, a reliable national energy data collection system, and inspection and enforcement mechanisms.

---

38 William Nordhaus, "The *Stern Review* on the Economics Of Climate Change," November 17, 2006, www.econ.yale.edu/~nordhaus/homepage/SternReviewD2.pdf.

39 Stern, *Economics of Climate Change,* chap. 16, note 28.

40 Many have proposed such an approach; see, for example, Richard B. Stewart and Jonathan B. Wiener, "Practical Climate Change Policy," *Issues in Science and Technology* (Winter 2003), www.issues.org/20.2/stewart.html. See also a book by the same authors, *Reconstructing Climate Policy: Beyond Kyoto* (Washington, D.C.: American Enterprise Institute, 2003).

Part VI of the Stern review addresses how collective action might be achieved. The review notes that even if developed countries reduce their emissions by 60 percent from 1990 levels by 2050, developing economies would need to reduce emissions by 25 percent from the 1990 level to achieve stable emissions that eventually would lead to atmospheric CO_2 equivalent concentrations of 550 ppm. If developed countries reduced their emissions by 90 percent, then developing economies could increase their emissions by 50 percent to achieve the same outcome.[41]

An alternative approach that has been suggested is that Annex 1 countries might impose a tax on imports from non-Annex 1 countries as a way of encouraging these countries to adopt carbon constraints. Presumably, the level of the import tax is set by some combination of environmental external cost attributed to greenhouse gas emissions and from the difference in energy costs in Annex 1 and non-Annex 1 countries. The press reports that Prime Minister Dominique de Villepin of France has made such a proposal.[42] I hope such an approach is not adopted. A coercive approach to this convergence problem is unlikely to be practical or to influence the nations that are the most obvious targets—the United States and China.

A realistic appraisal is that we are making no progress in achieving such ambitious goals, and that it is inconceivable that any significant progress will be made if the United States, currently the world's largest emitter of greenhouse gases, is not an active participant in the process. From my perspective, the greatest political danger in the United States is not continued adherence to the position of the current administration that global warming does not require collective action. Rather, the danger is that the Congress promptly passes a weakened version of the McCain-Lieberman bill, which puts in place a relatively restricted cap-and-trade system with an opt-out feature permitting open-ended

41 Stern, *The Economics of Climate Change,* 459, Figures 21.1 and 21.2.

42 "French Plan Would Tax Imports from Non-Signers of Kyoto Pact," *New York Times* (Reuters), November 14, 2006; see also the Web site of the prime minister: www.premierministre.gouv.fr/en/information/latest_news_97/sustainable_development_unveiling_the_57272.htmlministre.gouv.fr/en/information/latest_news_97/sustainable_development_unveiling_the_57272.html.

purchase of allowances at a low price.[43] Initial constraints of this sort are insufficient to cause any change and are considered as a first step toward meaningful constraints. The risk is that Congress agrees to adopt the low-cost first step, believes the problem to be "solved," and delays serious deliberation about more stringent controls.

Consequences and Choices

My view is that four changes are needed to make progress on the global warming issue.

First, the United States must adopt a carbon emission control policy.

Second, an agreed framework is needed between developed economies and large emerging economies about how the costs of carbon emission control will be shared.

Third, the leading technology for controlling greenhouse gas emissions is CCS.[44] Trilateral countries should urgently launch five to six large CCS projections around the world in order to demonstrate the technical feasibility and public acceptance of carbon sequestration.

Fourth, there should be expanded use of nuclear power; this is discussed in the next section.

Because there has been so little progress on reaching a workable approach to controlling greenhouse gas emissions, the question naturally arises: What happens if countries take no action?

First is to hope that the effects of greenhouse gases on climate occur sufficiently slowly that there is time for world societies and economies to adapt to the change gradually and without great suffering and disruption. And there is always the possibility of an unexpected tech-

---

43 For an analysis of the McCain-Lieberman bill, S-139, see William A. Pizer and Raymond J. Kopp, "Summary and Analysis of McCain-Lieberman—'Climate Stewardship Act of 2003'; S.139, introduced 01/09/03," Resources for the Future, Washington, D.C., January 28, 2003, www.rff.org/rff/Core/Research_Topics/Air/McCainLieberman/loader.cfm?url=/commonspot/security/getfile.cfm&PageID=4452. The Energy Information Administration, Office of Integrated Analysis and Forecasting, has posted an analysis of the original version of the 2003 McCain-Lieberman bill, "Analysis of S.139, the Climate Stewardship Act of 2003: Highlights and Summary," June 2003, www.eia.doe.gov/oiaf/servicerpt/ml/pdf/summary.pdf.

44 For a review, see Metz et al., eds., *IPCC Special Report on Carbon Dioxide Capture and Storage.*

nical innovation that will greatly alleviate the problem. Thomas Schelling is a respected and knowledgeable proponent of this view.[45]

The second possibility is that a disruptive and costly climate event will occur and that this event will convince the public and political leaders that action is urgently needed. I am skeptical that crisis is a good catalyst for adopting wise policy. Some suggestions for possible crisis responses reinforce my conviction.

If a climate crisis occurs, the mitigation approach may not work because it may then be too late for these measures to take effect. Attention will turn to the possibility of "geotechnical" solutions, which refer to active human intervention intended to reverse the effects on global climate of the emission of greenhouse gases from human activity. Recently, an increasing number of experts have been suggesting that more serious consideration be given to geotechnical solutions because emission mitigation does not seem to be acceptable. One prominent person to advance recently this possibility was Nobel Laureate Paul J. Crutzen.[46]

I do not wish to test the patience of those who are not technically knowledgeable but nevertheless experienced and thus properly cautious about accepting big risks. So I simply list some of the geotechnical measures that are under discussion to give an impression of the active measures that may be possible to counterbalance the global warming effects of greenhouse gas emissions:[47]

- Adding aerosols to the stratosphere (sulfate, soot, dust, metallic particles);
- Placing balloons or mirrors in the stratosphere;

---

45 Thomas C. Schelling, "Some Economics of Global Warming," *American Economic Review* 82, 1 (1992); "The Cost of Combating Global Warming," *Foreign Affairs,* November–December 1997; "What Makes Greenhouse Sense," *Foreign Affairs,* May–June 2002.

46 Paul J. Crutzen, "Albedo Enhancement by Stratospheric Sulfur Injections: A Contribution to Resolve a Policy Dilemma?" *Climatic Change* 77, no. 3–4 (August 2006): 211–220.

47 A number of these measure were discussed at an informal meeting at a NASA, Ames Research Center–Carnegie Institution workshop, November 18–19, 2006. For background, see D. W. Keith, "Geoengineering the Climate: History and Prospect," *Annual Review of Energy and the Environment* 25 (2000): 245–284.

- Making deserts more reflective;
- Modifying the ocean albedo;
- Fertilizing the ocean to increase CO_2 uptake; and
- Creating high-altitude nuclear explosions to induce a nuclear "spring."

These measures do not have the benign character of reforestation. Instead they involve human intervention at a large scale in order to correct the global climate system for the perturbation in the global climate system caused by anthropomorphic greenhouse gas emissions. It is a very tall technical order to demonstrate control over such intervention with the level of confidence demanded by responsible public action. The geotechnical intervention option's greatest value is that its prospects should encourage Trilateral countries to redouble their efforts to seek a mitigation solution.

Nuclear Energy

Concern with global warming and the increasing price of natural gas has understandably stimulated interest in expanded use of nuclear power. In the short run, nuclear power can substitute for coal and natural gas–fired electricity generation; in the long run, the possibility exists of using electricity to a greater extent in both individual and mass transportation systems.

At present, nuclear power generation accounts for about 16 percent of the world's electricity production; most of the installed nuclear capacity is in the United States, Europe, and East Asia. But nuclear power use faces considerable challenges: capital costs must go down; progress must be made on waste management; and best safety practices in design, construction, and operation must be assured. Most important, for security, any expansion of commercial nuclear power must not be an avenue for countries to move toward or acquire a weapons capability. The example of Iran is the most immediate.

The nonproliferation issue is especially critical because projections of growth of electricity consumption in emerging economies are two to three times the rate projected for the developed world. We are speaking about countries such as Indonesia, Turkey, Egypt, Taiwan, and South Korea that might raise some concerns, but also countries such as Chile, Argentina, and Brazil.

The 2003 MIT study, *The Future of Nuclear Power,*[48] suggested that a tripling of global nuclear generating capacity, from about 300 GWe[49] to 1,000 GWe by mid-century, might be feasible. However, nuclear power in the developing world would grow from about 10 GWe in 2000 to 300 GWe in 2050. The objective of Trilateral countries should be to assure that these countries have access to the benefits of power generation without providing them with the technologies that invite the spread of nuclear weapons.

The most potentially dangerous technologies are enrichment (at the front end of the fuel cycle) and reprocessing (at the back end of the fuel cycle); see figure 11. Enrichment lifts the percentage of the fissile uranium-235 from 0.7 percent in natural abundance to about 3 percent for commercial fuel. Enrichment technologies such as centrifuges can be used to increase the percentage of U-235 to weapons grade. Reprocessing refers to chemical separation of the actinides, uranium and plutonium, from the fission production in the spent fuel. The plutonium isotope, Pu-239, formed during reactor operation by neutron absorption of the plentiful uranium isotope, U-238, is directly bomb usable.

Trilateral countries acting through the Group of Eight (G-8) have acknowledged the need to adopt new mechanisms to improve the proliferation resistance of commercial nuclear power. The objective is to adopt new proliferation-resistant mechanisms that would make it more difficult to use nuclear power as a path to acquiring nuclear weapons capability, thereby reducing the risk of proliferation.

It is unrealistic to believe, however, that adopting new mechanisms is a guarantee that a state will not be able to acquire nuclear weapons. Iran, a Nuclear Non-Proliferation Treaty (NPT) signatory, is an example of a state that appears to be seeking both nuclear power and nuclear weapons technology. States that seek a nuclear weapons capability do so because of their perception of their security interests, and when they do so, if history is any guide, such states will pursue a clandestine route to acquiring the strategic nuclear material needed for the bomb. Adopting new mechanisms to safeguard nuclear power from diversion seeks to preserve this distinction.

---

48 S. Ansolabehere et al., *The Future of Nuclear Power: An Interdisciplinary MIT Study* (Cambridge: Massachusetts Institute of Technology, 2003), http://web.mit.edu/nuclearpower/.

49 A large nuclear power station has a capacity of about 1,000 megawatts electric output (MWe), which equals 1 gigawatt (GWe).

Figure 11. The Nuclear Fuel Cycle

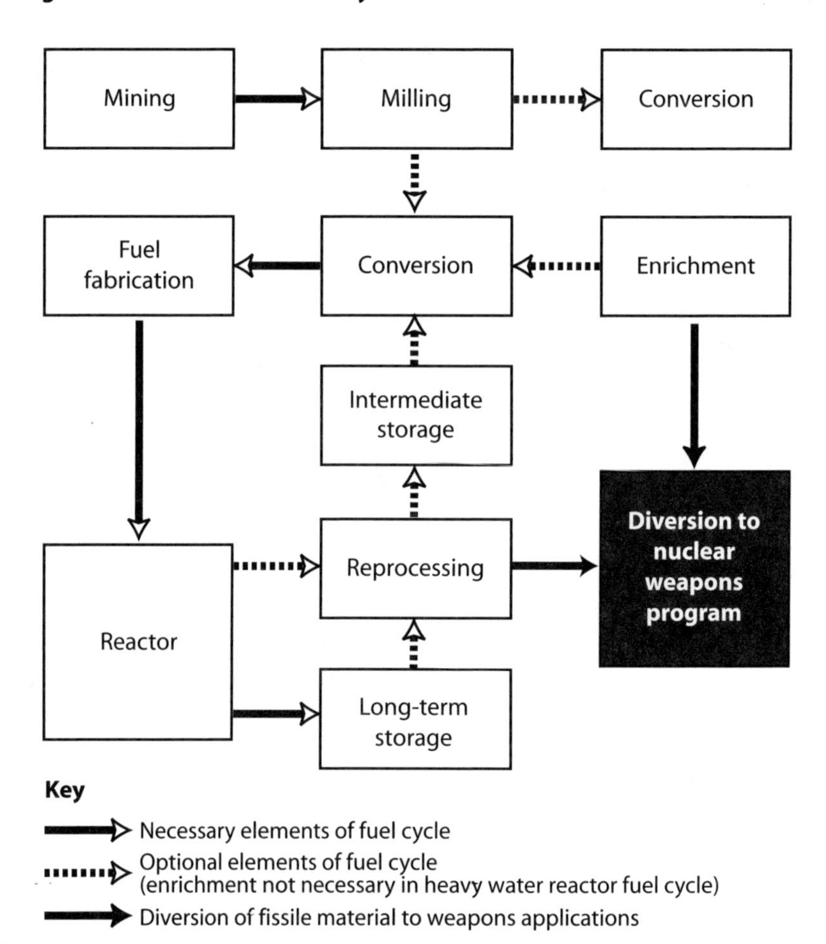

Source: "The Nuclear Fuel Cycle," Center for Nonproliferation Studies, Monterey, Calif., http://cns.miis.edu/research/wmdme/flow/iran/index.htm#1.

Trilateral countries and the G-8 have adopted a common approach to strengthening proliferation safeguards in anticipation of a possible expansion in nuclear power use. The G-8 has taken a series of measures beginning in its meeting in 2004 at Sea Island, Georgia, then in 2005 at Gleneagles, Scotland, and culminating in 2006 in St. Petersburg, Russia, where the G-8 announced its support of new mechanisms for nuclear supplier states to supply fuel cycle services to states that want to use nuclear power.[50]

Here is how it might work:[51]

Countries that do not possess uranium enrichment and plutonium reprocessing facilities would agree not to obtain any such facilities and related technologies and materials. In exchange, they would receive guaranteed cradle-to-grave fuel services under an agreement that was financially attractive and signed by all those countries in a position to provide them. The International Atomic Energy Agency (IAEA) would sign also and would apply safeguards to any such fuel cycle activities covered by the agreement in addition to its traditional safeguard activities with regard to the reactors in the recipient states. The idea is to make costly indigenous fuel cycle facilities less attractive than reliable fuel cycle services from a few nuclear supplier states.

To be effective, the arrangement would need to include a guarantee to the recipient country at least for enrichment services. It is not likely that a recipient country would make an investment in a reactor without an international guarantee that a quarrelsome U.S. Congress could not abruptly terminate commercial contracts for enriched fuel, for example. The guarantee would be strengthened by an international enrichment "bank" or "reserve," operated perhaps by the IAEA. In a remarkable display of public support for such a security initiative, former U.S. senator Sam Nunn, the cochair of the U.S.-based Nuclear Threat Initiative, announced that Warren Buffett had pledged $50 million toward a total of $150 million for a low enriched uranium stockpile owned and operated by the IAEA.[52]

This is a new approach that amounts to an important revision of the terms of President Dwight D. Eisenhower's 1953 "atoms for peace" deal between nuclear have and have-not states. The implicit deal in "atoms for peace" was that nuclear weapon states would provide ac-

---

50 "Global Energy Security," G-8 Summit 2006, St. Petersburg, Russia, July 16, 2006, item no. 31, http://en.g8russia.ru/docs/11.html.

51 John Deutch, Arnold Kanter, Ernest Moniz, and Daniel Poneman, "Making the World Safe for Nuclear Energy," *Survival* 46, no. 4 (Winter 2004–2005).

52 See former senator and NTI cochair Sam Nunn's speech and NTI's press release, September 19, 2006: www.nti.org/c_press/speech_Nunn_IAEAFuelBank_FINALlogo.pdf; http://nti.org/c_press/release_IAEA_fuelbank_091906.pdf.

cess to technology and nuclear material to non-weapon states in exchange for the recipient non-weapon state agreeing to forgo nuclear weapons. The proposed conditions for today's deal are more stringent: recipient states get access to technology and nuclear power reactors but not to the more sensitive parts of the fuel cycle. These fuel cycle services—enrichment and reprocessing—would come from a restricted number of nuclear supplier states. The universe of nuclear supplier states remains to be defined; presumably, supplier states initially would consist of the nuclear weapon states plus some others, for example, Germany, Japan.

Of course, a good policy idea may be a long way from a functioning policy. As yet there is no concrete example of such an arrangement. At one point, the Iranian Bushehr-1 reactor, constructed by the Russians, seemed to be a good model. The Russians would "lease" the fuel, that is, they would provide enriched fuel and take back the depleted fuel for reprocessing, or disposal, or both. However, the unwillingness of the Iranians to suspend their enrichment activities at Natanz has for the time being halted any progress on Trilateral support for Iran's nuclear program.[53]

Brazil's plan for construction of a $210 million enrichment facility at Resende (a project run by the Brazilian navy) presented a second opportunity to achieve this new fuel cycle arrangement.[54] The United States chose to acquiesce to the Brazilian enrichment program and to object to the Iranian enrichment program on the grounds that the latter was dangerous and the former was not. I do not believe a proliferation policy will be workable in the long run unless the rules have international scope and can be consistently and objectively applied. The Brazilian decision to proceed with its domestic enrichment plant is a setback to the proposed G-8 policy to internationalize fuel cycle services.

---

53 For a description of Iran's nuclear activity, see "Iran Profile," NTI, Washington, D.C., http://nti.org/e_research/profiles/Iran/index.html.

54 For background, see "Brazil Profile," NTI, Washington, D.C., http://nti.org/e_research/profiles/Brazil/index.html; see also Sharon Squassoni and David Fite, "Brazil as a Litmus Test: Resende and Restrictions on Uranium Enrichment," *Arms Control Today*, October 2005, www.armscontrol.org/act/2005_10/Oct-Brazil.asp.

A third event was the December 18, 2006, signing of the U.S.-India Peaceful Atomic Energy Cooperation Act.[55] While the strengthening of the political relationship between the United States and India has much to recommend it, this action is not positive on nonproliferation grounds because it serves to legitimize a country that is not a NPT signatory; India is instead an "undeclared" nuclear weapons state that does not permit IAEA inspections of its nuclear facilities.

Practical realization of the new policy calls urgently for one example that works. An important, infrequently acknowledged barrier is that few countries are willing to accept the return of spent fuel. Russia will take back fuel of Russian origin; the situation with regard to the European Union and France, in particular, is less clear. I say with some confidence that the United States will not be willing to accept returned spent fuel until its waste management program is in better health.

I conclude with some emphasis that all Trilateral countries should support the G-8 nonproliferation initiatives that seek to provide assurances of fresh fuel and spent fuel management to states that agree not to pursue enrichment and reprocessing programs.

Advanced Fuel Cycle Development

In parallel with the nonproliferation initiative, the G-8 announced at St. Petersburg, Russia, its support for the development of "innovative nuclear power systems."[56] Several members of the G-8, notably France, Japan, and Russia, are eager to support the U.S. initiative, the Global Nuclear Energy Partnership (GNEP), to develop a new advanced fuel cycle.[57] The GNEP advanced fuel cycle will include a new separation

---

55 For a description of the provisions of the act, see the fact sheet from the Office of the White House Press Secretary, December 16, 2006, www.whitehouse.gov/news/releases/2006/12/20061218-2.html.

56 "Global Energy Security," G-8 Summit 2006, item no. 29, states:

> The development of innovative nuclear power systems is considered an important element for efficient and safe nuclear energy development. In this respect, we acknowledge the efforts made in the complementary frameworks of the INPRO project and the Generation IV International Forum. Until advanced systems are in place, appropriate interim solutions could be pursued to address back-end fuel cycle issues in accordance with national choices and nonproliferation objectives.

57 A description of the GNEP can be found on the U.S. Department of Energy Web site, www.gnep.energy.gov/.

process that is somewhat more proliferation resistant (because the plutonium and uranium are not separated into separate streams as occurs in the conventional PUREX separation process) and a new family of reactors, Generation IV, which are capable of burning the long-lived radioactive actinide isotopes.

The GNEP intends that the United States return to a "closed" nuclear fuel cycle, where plutonium is recycled and mixed with uranium to form mixed oxide fuel to produce power in the reactor. The United States presently uses an "open" fuel cycle, where the spent fuel from reactors is not recycled but discarded in a geologic repository with its long-lived actinide isotopes. The alleged advantages of the closed cycle are: (1) it makes the waste management task easier because the absence of long-lived isotopes means the nuclear waste's radioactivity decays sooner (after several hundred years, rather than several tens of thousand years); (2) the uranium resource base is extended manyfold because of the breeding and reuse of plutonium in power reactors; and (3) the new system can be made proliferation resistant.

The United States had a long debate in the 1970s about the relative merits of a closed versus open fuel cycle. President Ford cancelled plans for reprocessing of commercial spent fuel in 1975. President Carter placed the country on the open fuel cycle path, canceled projects related to the closed fuel cycle—for example, the Clinch River breeder reactor—and argued in international forums such as the International Fuel Cycle Evaluation study that the open fuel cycle, as then configured, presented unacceptable risks in international commerce because it made bomb-usable separated plutonium widely available. The U.S. view influenced the attitudes of Trilateral countries about the proliferation risks of fuel cycle exports but did not convince Trilateral countries to abandon the closed cycle indigenously. For example, France is the world leader in the development and operation of commercial nuclear fuel reprocessing at La Hague. Japan and the United Kingdom have had mixed results in their reprocessing efforts.

Here I want to stress that it is by no means clear that the United States will proceed with the GNEP. There is strong criticism of this approach, which I share, and for the program to succeed, it will require decades of work and the support of many administrations.[58] The new GNEP architecture does not offer a clear advantage for waste disposal because the long-term environmental benefits of removing actinides from the waste must be balanced against the near-term risks of operating complex reprocessing and fuel fabrication plants. All agree that the closed cycle will be more expensive (although not a large percentage of the total cost of electricity) than the open cycle for many decades until, if, and when there is sufficient deployment of conventional nuclear power plants to drive up the cost of natural uranium ore to the point where reprocessing and fabrication of spent fuel is economic.

The strongest objection concerns the nonproliferation consequences of this strategy. At the same time that the G-8 is attempting to convince other countries not to deploy indigenous reprocessing technology, it announced that the G-8 should urgently pursue such technology for themselves. Other countries, such as Iran, Brazil, Turkey, South Korea, and Taiwan, might well wonder whether they are being asked to give up an important technology. The alleged improved proliferation resistance of the advanced reprocessing technologies is a fantasy. Supplier states such as France, Russia, and the United States are not proliferation risks. Essentially, other nations will not have a nuclear power industry of sufficient size to justify the large and expensive fuel cycle envisioned by the GNEP. In any case, if a country decided to divert material, relatively modest additional processing would be needed to recover pure plutonium.

My conclusion is that a major United States or G-8 effort to develop an advanced closed fuel cycle, rather than meeting the objective of "deploying advanced, proliferation resistant nuclear energy systems that avoid separation of pure plutonium and make it as difficult as possible to misuse or divert nuclear materials to weapons,"[59] will in fact derail the prospects for an orderly expansion of nuclear power throughout the world at a time when there are few alternatives to further emissions of CO_2 from electricity generating technologies.

---

58 For a thorough critique, see Richard K. Lester, "New Nukes," *Issues in Science and Technology,* Summer 2006, www.issues.org/issues/22.4/index.html; see also John Deutch and Ernest J. Moniz, "A Plan for Nuclear Waste," *Washington Post,* January 30, 2006, Sec. A.

59 Dennis Spurgeon, assistant secretary for nuclear energy, U.S. Department of Energy, "Assurances of Nuclear Supply and Nonproliferation" (speech at IAEA, Vienna, Austria, September 19, 2006), http://energy.gov/news/4173.htm.

In time, if use of nuclear power significantly expands around the world, it may be justifiable to adopt a closed fuel cycle. But, at present, pursuing the GNEP risks making that future less likely. As Matthew Bunn so aptly puts it, the GNEP substitutes the U.S. message in place since 1975, "we believe reprocessing is unnecessary and we are not doing it," with the message "reprocessing is essential for the future of nuclear power, but we will keep the technology away from all but a few states."[60]

The political dynamic in the United States that makes the GNEP attractive is the false hope that a closed cycle will lead to a waste disposal solution. The U.S. Congress is weary of cost overruns and delays in constructing the waste repository at Yucca Mountain, Nevada. Congress has the false hope that a closed cycle will be easier to accomplish and politically more acceptable. But opposition will mount quickly. Proliferation risk is a powerful public issue, and the GNEP supporters do not have an extensive political base compared with, for example, the farm interests that support corn-based gasohol. The GNEP will cost billions of incremental research and development dollars that many will judge could be better spent on other energy programs, especially on developing technologies that encourage greater energy efficiency. In sum, the future of the GNEP in the United States is highly dubious.

Encouraging Nuclear Power

What should Trilateral countries do to encourage nuclear power? First, work out effective procedures to assure international guarantees for nuclear fuel supplies; second, encourage supplier states, such as Russia, that are willing to take back spent fuel; third, strengthen the IAEA inspection regime, in particular by encouraging adherence to the IAEA "additional protocol,"[61] which gives the IAEA greater authority to chal-

---

60 Matthew Bunn, "Assessing the Benefits, Costs, and Risks of Near-Term Reprocessing and Alternatives" (testimony before the Subcommittee on Energy and Water Appropriations, U.S. Senate, September 14, 2006), http://bcsia.ksg.harvard.edu/BCSIA_content/documents/bunn_gnep_testimony.pdf.

61 For additional information, see "Model Protocol Additional to the Agreement(s) between State(s) and the International Atomic Energy Agency for the Application of Safeguards," Document INFCIRC/540, International Atomic Energy Agency, Vienna, Austria, September 1997, www.iaea.org/Publications/Documents/Infcircs/1998/infcirc540corrected.pdf.

lenge inspections of undeclared facilities in NPT party states; and fourth, establish a consortium among nuclear supplier nations with existing technologies and financial instruments able to offer developing nations nuclear power at reasonable cost and without proliferation risk. This amounts to an open cycle strategy for nuclear exports for at least the next several decades.

Of course, a research program on advanced nuclear fuel cycles should go forward. But these research activities should be limited to laboratory-scale research on new separation technologies, design, analysis, and simulation of new fuel cycle systems as well as experiments to obtain engineering data. Large demonstration facilities and any suggestion of near-term deployment of a closed fuel cycle system should be deferred for the foreseeable future.

Conclusions

I have discussed four energy security issues. Here are summary conclusions about what should be done about each:

1. To mitigate the effects of oil and gas import dependence, we must begin the process of a transition away from a petroleum-based economy and recognize the inevitable dependence on petroleum until that transition is accomplished.
2. Reducing the growing vulnerability of the energy infrastructure calls for greater cooperation for Trilateral countries and others involved in international energy markets.
3. Both developed and developing economies need to curb CO_2 and other greenhouse gas emissions to avoid the adverse consequences of climate change or face the prospect of active engineering of the globe's climate.
4. The need for encouraging expanded use of nuclear power means that new measures must be adopted to reduce the increase in proliferation risk that would result from the spread of dangerous fuel cycle services—enrichment and reprocessing.

We justifiably should be concerned that the world is not making sufficient progress on these issues. One possibility is that the world will continue to muddle along and make the inevitable adjustments. Another possibility is that a severe crisis will change the attitude of the public and its leaders about what needs to be done. I am uncomfortable with either of these possibilities because I believe each will involve much higher economic and social cost than is necessary. A much better option is to manage the significant social, technical, and economic aspects of the energy transitions the world will undergo. I hope that the Trilateral Commission, both as an organization and as individuals, will strive to make progress on these energy issues in the years ahead, appreciating that energy and security issues are not divisible, and I look forward to a promising assessment at future meetings.

2

Energy Security and Climate Change: A European View

Anne Lauvergeon

At the same time as the population of the planet has increased, the globalization of the economy and the growth to which this has led have contributed to intensified industrialization and urbanization. Globalization and growth have raised the demand for mobility and transport. In a more general sense, they have added to the world's energy needs. Air pollution is only one aspect of the damaging consequences of this development. In addition to the risk of interruptions to energy supplies, there is now the threat to the global climate through the intensive use of carbon fuels and greenhouse gas emissions.

How can sustainable growth be maintained with sufficient energy supplies, yet without damaging the whole planet? At the request of the Group of Eight (G-8) heads of state, the International Energy Agency (IEA) looked at possible ways of developing a clean, clever, and competitive energy future. Its conclusions were alarming:

The energy future which we are creating is unsustainable. If we continue as before, the energy supply to meet the needs of the world economy over the next twenty-five years is too vulnerable to failure arising from under-investment, environmental catastrophe or sudden supply interruption.[1]

These questions are not being addressed in the same manner and with the same level of intensity in the various countries of the Trilateral Commission. In the United States, the debate on energy security

---

This paper was prepared with Jean-Pol Poncelet, European associate author and adviser to Anne Lauvergeon. Mr. Poncelet is senior vice president of Areva, Paris, and former deputy prime minister, minister of defense, and minister of energy of Belgium.

1 *World Energy Outlook 2006* (Paris: International Energy Agency, 2006).

seems to focus on the relationship between the economic activity that is developing in the domestic and international energy markets and the responses in terms of foreign policy.[2] The intensity of energy consumption in the United States, significantly higher than in all the other Trilateral countries, does indeed have major economic and political consequences. The latter are exacerbated by the growing U.S. dependence on crude oil imports, even though its overall energy dependency is slightly lower than Europe's.[3]

For their part, Europeans do not currently share a common energy policy, even though historically the original constitutive treaties (the CECA Treaty that set up the European Coal and Steel Community and the Euratom Treaty that set up the European Atomic Energy Commission) were energy oriented: Regarding energy, the European Union (EU) member states essentially operate on the basis of a national prerogative.[4] The European Commission (EC) nonetheless considers that EU-wide action is now essential in this area.[5] The energy issue has been more broadly addressed by the EC in the security of a supply–competitiveness–climate change triangle. The three problems have been linked globally by the EC. To begin with, there is the dependence on imported energy, which could increase from 50 to 65 percent by 2030. Then there is the constraint of rising energy prices, which increase the net transfer of wealth, adversely affecting competitiveness and employment. Finally, there are the challenges of climate change in relation to

---

2 See John Deutch, "Priority Energy Security Issues," in this volume.

3 Total energy consumption (primary energy, 2004) for the United States amounted to 7.8 metric tons of oil equivalent per inhabitant (toe/capita) and 3.7 toe/capita in Europe (EUR-25). For crude oil, these values were 3.1 and 1.5 tons per inhabitant (t/capita). For gas, they were 2.2 and 1.1 thousands of cubic meters per inhabitant (km3/capita). The European Union's overall energy dependence (for 2005) was approximately 56 percent, while that of the United States was 31 percent and that for North America as a whole, 17 percent.

4 The draft Treaty for a European Constitution (July 2003) that is in the process of approval by member states proposes the allocation of a shared competence in energy matters to the EU (article 13, section 2). It was rejected by referendum in two member states.

5 "Une politique de l'énergie pour l'Europe" (communication de la Commission au Conseil Européen et au Parlement Européen, COM [2007] 1 final, Bruxelles, 2007).

the commitments made by the EU at Kyoto. The conclusions of this analysis are no less negative: "Current energy policy cannot be maintained in the context of sustainable development."

Unlike the United States, the European Union is already committed to reducing its carbon emissions. It is doing so through the application of the 1994 UN Framework Convention on Climate Change and the 1997 Kyoto Protocol. The EU target is to limit the rise in average temperature of the planet (compared with that of the preindustrial era) to 2°C. To this end, the EC has proposed an energy policy to the EU, based on the unilateral commitment to achieve, whatever the circumstances, a reduction of at least 20 percent in greenhouse gas emissions by 2020 compared with 1990. In parallel, the EC has restated the importance of joint, united action on a worldwide scale. In the event that the overall commitment made elsewhere would lead to reduced worldwide emissions by 50 percent compared with 1990 levels before 2050, the EC has suggested that Europe should set itself the more demanding target of a 30 percent reduction by 2020 instead of 20 percent.

Ensuring sustainable growth while observing such restrictive targets and simultaneously providing for the energy required for development involves the combination of three types of initiatives: first, a significant rise in energy efficiency; next, a demonstration of the feasibility and the widespread implementation of the technologies for the capture and storage of carbon emitted in the burning of fossil fuels; and, last, the development of low-carbon energies, that is, renewables and nuclear.

The timescale relating to the availability and the industrial implementation of these solutions is highly variable. The use of many technologies to improve energy efficiency could spread quite rapidly, for the most part with no other incentive than the savings that result. Furthermore, recent experience has shown that the use of several forms of renewable energy could easily be increased through a suitable tax policy or pricing policy or both. In most Trilateral countries, including those with deregulated markets, nuclear power is now competitive for the production of base-load electricity without special involvement by the public authorities. Conversely, it is not currently possible to draw up a clear forecast of the feasibility for operational carbon dioxide capture and storage systems for coal or gas power stations. Furthermore, the benefits of fuel cell and hydrogen technologies for private transport and the decentralized production of energy will, in all likelihood, not be possible to exploit in the near future.

This paper addresses the objective set out by the Trilateral Commission, which wished to look at energy security from the particular angle of the problems of hydrocarbon supply, the vulnerability of energy infrastructure, climate change, and nuclear energy. The notable commitment by the EU to the Kyoto Protocol and the more general concern of European citizens with regard to climate change, the importance of nuclear energy in the production of energy by EU member states, and European leadership in industrial nuclear activities—throughout the entire nuclear fuel cycle—require some of these problems to be looked at more specifically. There thus follow some considerations that illustrate more clearly European "sensitivity" to these issues. They draw inspiration from the concerns and experience of an industrial group active in the energy sector, committed to sustainable development, and aware of its responsibilities to society. It has added its findings to the analysis carried out elsewhere on the same subject.

Climate Change Issue

> There is irrefutable scientific evidence that demonstrates the need to take urgent action to fight climate change. [The costs of inaction] are not only economic, but also social and environmental, and these will be borne primarily by the poor, in both developing and developed countries. The absence of a reaction will have serious consequences for energy security, both at local and worldwide level. Most of the solutions are already known, but governments have to adopt them immediately in order to put them into practice.[6]

All countries—and in particular those of the Trilateral Commission—are affected by climate change. On the one hand, the developed countries are responsible for the majority of the industrially derived greenhouse gases that have built up in the atmosphere, and they have

---

6 "Limiter le réchauffement de la planète à 2°C. Route à suivre à l'horizon 2020 et au-delà" (communication de la Commission au Conseil, au Parlement Européen, au Comité Economique et Social Européen et au Comité des Régions, COM [2007] 2 final, Bruxelles, 2007).

the technological and financial capacity to act. But, on the other hand, given the growth in the economies and emissions of certain developing countries—in both relative and absolute terms—their contribution to the phenomenon will rapidly exceed that of the former.[7] The result of this is that even the most draconian measures that might be taken by developed countries will not only be less effective, they might also be insufficient. The United Nations Framework Convention on Climate Change (UNFCCC) has clearly identified this differentiation of responsibilities. This is why developed countries have been invited to take the first step toward limiting or reducing their emissions through the application of the Kyoto Protocol.

Although all Trilateral countries are affected, and even if the majority claim to be concerned, there nevertheless exist differences of opinion among them on the subject of climate change: sometimes on the diagnosis, cause, and scale of the phenomenon; most often on the strategy, timescale, and resources required to confront it. Debating these problems and seeking a consensus against a Trilateral background would thus be advisable, especially with the prospect of the forthcoming "post-Kyoto" negotiations, in other words, negotiations on the framework to be put in place after 2012: "I'm talking about climate change . . . because there is perhaps no other issue where we agree so much, yet understand each other so little."[8]

Reality of Climate Change

Experts worldwide recently brought to the attention of decision makers a summary of knowledge on climate change, including the results of the last six years of research.[9]

---

7 According to *World Energy Outlook 2006*, greenhouse gas emissions from China could exceed those of the United States, the world's current leading emitter, by the end of the decade. And worldwide, by the same date, carbon dioxide emissions from non-OECD countries could be greater than those from the OECD zone.

8 Kurt Volker, principal deputy assistant secretary for European and Eurasian affairs, U.S. Department of State (speech at the German Marshall Fund, Berlin, Germany, February 12, 2007).

9 "Climate Change 2007: The Physical Science Basis, Summary for Policymakers," Contribution of Working Group I to the Fourth Assessment Report of the Intergovernmental Panel on Climate Change (IPCC Secretariat, Paris, February 14, 2007), www.ipcc.ch/SPM2feb07.pdf.

Worldwide concentrations of several greenhouse gases grew following human activities, and they currently greatly exceed preindustrial values. The gases involved are carbon dioxide (CO_2), which results from the burning of fossil fuels and changes in land use, and methane (CH_4) and nitrous oxide (N_2O), which are mainly due to agriculture.

The resulting warming of the climatic system is unequivocal: It is obvious from the observations of increases in worldwide average sea and air temperatures, widespread melting of snow and ice, and the rise in average sea level around the world. At the level of continents, regions, and oceanic basins, many long-term changes in climate have been observed. These concern temperatures and Arctic ice, the quantity of precipitation, ocean salinity, the structure of winds, and certain extreme meteorological situations such as droughts, heavy precipitation, and heat waves.

The warming of the last half century is atypical for at least the last 1,300 years. Approximately 125,000 years ago—the last time that the polar regions were significantly warmer than they are now for a long period—the reduction in volume of the polar ice caps led to a rise in sea level of four to six meters. Most of the rise in average worldwide temperatures seen since the middle of the twentieth century is almost certainly due to the observed increase in man-made greenhouse gases. Human influence can also be noted on other aspects of climate such as ocean warming, average continental temperature, extreme temperatures, and wind structure.

Analysis of climate models made using the findings of these observations enable a likely range to be identified for climate sensitivity.[10] Even if concentrations of all greenhouse gases and aerosols had remained constant at 2000 levels, a warming of 0.1°C per decade would have resulted. The continuation of greenhouse gas emissions at current levels or above would cause additional warming and lead to many changes to the climate worldwide throughout the twenty-first century, which would almost certainly be greater than those witnessed during the twentieth century. The increase in sea temperature and level owing to man's activities would almost certainly continue for centuries be-

10 Without new measures to limit greenhouse gas emissions, the world will probably warm by an additional 1.8°C–4°C over the course of this century after having increased by 0.7°C during the past one.

cause of the timescale associated with climate processes and feedback, even if greenhouse gas concentrations were stabilized.

These phenomena would alter the operation of ecosystems, the distribution and availability of water resources, food production, and the occurrence of storms and droughts.[11] Their environmental, social, and economic impact on generations to come—even if their scale and development over time may be subject to debate because these are difficult to assess accurately—could be considerable. And any delay in taking the necessary action might increase the cost of any essential correction and prevention measures. At the request of the British government, Sir Nicholas Stern recently reviewed the economics of climate change.[12] He reports that, without further action to reduce emissions, the overall costs and risks of damage owing to climate change will reduce global gross domestic product (GDP) by at least 5 percent a year, and in the long term by possibly as much as 20 percent or more. In contrast, the review estimates that the cost of reducing emissions to avoid the worst impacts of climate change can be limited to around 1 percent of global GDP per year through use of the most cost-efficient instruments.[13]

Differentiated Approaches

Europe, which approved the Kyoto Protocol, is committed during the period 2008–2012 to reducing its greenhouse gas emissions by 8 per-

---

11 The EC estimates the number of people who died prematurely in the EU as a result of heat and increased air pollution during the summer 2003 heat wave at 20,000. Many experts believe this kind of summer could become the norm in Europe in a quarter century.

12 Nicholas Stern, *The Economics of Climate Change: The Stern Review* (Cambridge: Cambridge University Press, 2006), www.hm-treasury.gov.uk/independent_reviews/stern_review_economics_climate_change/stern_review_report.cfm.

13 In addition, it is worth noting that the European Commission indicates that EU action on climate change would considerably strengthen energy supply security within the EU (gas and crude oil imports would both fall by 20 percent by 2030 compared with the status quo). Furthermore, the EC estimates that a reduction of 10 percent in EU CO2 emissions between now and 2020, through its beneficial effect on air pollution, would generate health care savings of approximately 8 to 27 billion euros.

cent compared with 1990 levels. By setting targets and choosing to use economic instruments to internalize external costs, the EU wishes to provide the market with the opportunity to identify the most effective methods and to limit costs. By applying a cap-and-trade principle, it has established an emissions ceiling—initially for certain industries (power generators, cement manufacturers, steelmakers, chemists, paper producers), then gradually including other sectors such as transport—with a matching communitywide emissions-trading mechanism. According to its instigators, once the quota becomes a rare—and thus expensive—resource, industry would be encouraged to invest in clean technologies, creating incentives required to change the methods of energy production and consumption within Europe.[14] For this reason it could serve as the basis for international efforts to fight climate change. The EU does, however, recognize that at least for an intermediate period the competitiveness of some of these intensive energy-consuming industries could be compromised. The EC fears that some production would be relocated to other parts of the world with less severe environmental requirements. It proposes a number of ways to tackle this; for example, international limiting agreements for some industries, trade measures such as adjustments via border tax, and a special framework for targeted public funding. The EU is nonetheless aware that it is essential to reach an international agreement to limit climate change.

Unlike Europe, several developed countries, including the United States and Australia, did not ratify the Kyoto Protocol and do not seem to want to sign up to such a mechanism.[15] The current U.S. administration has indicated that it will not be making any initiatives toward ratification of the protocol because of its exemptions in favor of several

---

14 Since the system came into force, prices for carbon dioxide emissions quotas have collapsed. Quota allocations for the first phase of the carbon market were in fact too generous, and this combined with the effects of a mild winter and the relative weakness of the price of natural gas in Europe. For the next phase, 2008–12, the European Commission's firmness will be crucial. It will have to impose significantly lower ceilings on industry and member states if it wishes to increase the credibility of the system it established.

15 The situation in Canada, which ratified the Kyoto Protocol, became more complicated following contradictory statements by politicians on whether the country would achieve the targets and the institutional difficulties that might arise as a result.

major emitter countries such as China. The size of the U.S. economy, the biggest in the world, is sometimes used to justify the fact that the United States is also the world's largest greenhouse gas emitter, despite accounting for only 5 percent of the planet's population.

For the U.S. government, which says it wishes to make a distinction between the ends and the means, the Kyoto Protocol is just one means among many others. It thus wishes to stress not the targets resulting from regulation imposed by governments but voluntary efforts through technologies that enable emissions to be reduced without inhibiting economic growth. The United States therefore emphasizes the role of market mechanisms to encourage the development of cleaner energy technologies, making them economically competitive and thus making their use more widespread.[16]

Elsewhere, the United States is involved in a specific multilateral approach: The Asia Pacific Partnership on Clean Development and Climate (AP6) brings China, India, Japan, South Korea, and Australia together with the United States to tackle complementary energy, economic, and environmental goals. These countries are reported to account for 50 percent of the global population, 50 percent of the world economy, and 50 percent of all energy use. Under the AP6, they have identified almost 100 projects expected to deliver cumulative benefits of reduced greenhouse gas emissions, cleaner air quality, and reduced poverty levels through a growth-based approach. Initiatives contrary to the official position of the U.S. administration have, however, multiplied at the instigation of ten or so states and a number of cities that wish to pressure the federal government by applying the Kyoto Protocol principles themselves on a local basis. Several industrial groups have also called for action and recommended a prompt enactment of national legislation in the United States.[17]

---

16 These include, for example, the willingness to reduce vehicle fuel consumption by 20 percent in ten years, to increase the production and use of biofuels in cars, and, under the Clean Coal Initiative, to deploy advanced coal technology for cleaner and ultimately emissions-free energy, thanks to tax credits granted by the U.S. government.

17 *A Call for Action, Consensus Principles and Recommendations from the U.S. Climate Action Partnership: A Business and NGO Partnership* (Washington, D.C.: United States Climate Action Partnership, January 2007), www.us-cap.org/ClimateReport.pdf.

Last, the economies of several developing countries (in particular China, India, and Brazil) are growing rapidly and could account for half of all greenhouse gas emissions in twenty years' time, in parallel with the expected doubling of their GDP. It will be essential for them to slow the pace of growth of their emissions in order to reduce them. This need not involve restricting their development in a framework such as the Kyoto Protocol. Their emissions on a per capita basis could be taken into account, as well as the technical and financial feasibility of the measures envisaged to ensure the continuation of the economic growth and development of the countries concerned. The use of clean development mechanisms[18] could in particular be intensified thanks to an improvement in methods for financing international projects. But, in any event, action on the part of emerging countries will be essential to limiting global warming. This would be even more conceivable if developed countries—in particular, the largest of these, the United States—had themselves shown a willingness to act.

Time to Act

In the majority of the Trilateral countries, the media, public opinion, nongovernmental organizations, and, increasingly, the business world are becoming involved as a result of heightened awareness of the dangers of climate change. As the problem is global, it requires global solutions. Thus it is essential to win the support of all the major greenhouse gas–emitting countries for arrangements on emission limitations that are both technically possible and economically bearable. It is the duty of governments to draw up a series of consistent, stable, and effective long-term measures. They should combine political commitment—unilateral, if necessary—with market-based mechanisms, joint arrangements, and individual initiatives, addressing energy saving and efficiency, technology development, research and development (R&D) programs, the use of renewable energy, and controlled nuclear energy. Such objectives would require consensus within the Trilateral Commission on several subjects.

---

18 Clean development mechanisms generate "carbon credits" for investment in projects that lead to reductions in emissions in developing countries. These credits can then be used by developed countries toward their own targets. They should contribute to technology transfer.

International cooperation. We need to convince the United States on the one hand and China, India, and a number of additional countries on the other to join in discussions related to the post-2012 period, with a view toward building international arrangements on how to slow, stop, and reverse the growth of greenhouse gas emissions.

In particular, this requires a commitment from the United States to agree on, define, and implement policies relating to their own emissions, including a long-term CO_2 price. More broadly, such an objective assumes the involvement of emerging countries. The details of an interdependent system of development that contributes to emerging countries' growth—yet includes environmental constraints—needs to be defined with these countries. Transport-related problems can serve to illustrate the scale of the problem:[19] In developing countries, demand for transportation is increasing faster than population or GDP. If it is to be satisfied on the basis of the current energy model for developed countries (including urbanization and personal mobility), deadlock is guaranteed.

Technology and R&D. In the same way that Europe is proposing to broaden the use of regulatory tools, accompanied by targets and deadlines, U.S. industry and decision makers are rightly concentrating on the effectiveness of technological developments. Market-driven technological developments are indeed essential to increasing energy efficiency—in, for example, the field of cars and transport in general, housing, and industry—or they are essential to intensifying the use of new and renewable energies such as biomass, photovoltaic, wind, and hydrogen. There is thus vast potential for technological cooperation among Trilateral countries for the research, development, and demonstration of low-carbon energy technologies. The industrial world, which will play the most important role, must indicate the contributions it is prepared to make and the transitional measures that it expects for certain particularly exposed sectors. In the same way, the relative merits of various forms of government support under consideration should be discussed and debated, for example, incentives, public-private partnerships, and border-tax adjustments as well as the most suitable international trading mechanisms.

---

19 See Sylvie Kauffmann, "Comment croître sans polluer la planète?" *Le Monde* (Paris), February 13, 2007.

Nonetheless, we must avoid falling into the trap of believing in all-powerful technology producing ready-made solutions at the expense of genuine discussion about our way of life and relationship with energy: "Technology is the answer, but what's the question?"

An example of this is the expectation of integrated systems, including transport, capture, and storage of carbon emitted through the burning of fossil fuels (particularly coal). Such a system is too often presented as a miracle solution—proven and available, competitive and safe, socially acceptable and possible to impose everywhere immediately. This is far from the case. Demonstration projects are essential; some are already under way in Trilateral Commission countries. Approximately ten such projects are listed by the IEA. If successfully completed, they would enable, by 2015, barely 0.2 percent of expected coal-fired power station emissions to be avoided. Thus, for a half century this technology will not play a significant role in the limitation of carbon emissions. Therefore we must ensure a proper division between those technologies potentially available for reducing emissions over the next quarter century and those that are more distant and might contribute to a fundamental change in the energy system.

Nuclear energy. In many Trilateral Commission countries, nuclear energy continues to divide public and governmental opinion. It is essential that governments continue down the path of transparency, education, and conviction. The following aims at contributing to this process.

The Case for Nuclear Power

Because it produces practically no greenhouse gases, nuclear power is an option for reducing carbon dioxide emissions. Nuclear technology is able to satisfy the base-load electricity needs of industry, public transport, and many household applications all the better because it reduces dependence on imported fossil fuels. It is also a means of providing substitute energy to make up for the intermittent generation of most renewable power sources. The R&D work under way internationally will make it possible to build even safer new reactors that will make better use of the energy potential of uranium, thereby greatly reducing the volume and residual toxicity of waste. Furthermore, the need for new and alternative fuels for transport (based on gasification, coal to liquid, and biomass) will increase the demand for hydrogenous products that could best be produced by large-scale, nuclear-based platforms combining electricity, steam and hydrogen output, as well as by electrolysis systems.

Nuclear technology is proven, competitive, and safe; and nuclear energy has largely predictable generation costs. In most countries it is the only energy source to account for its long-term external costs (insurance, dismantling, waste processing, and disposal). New reactors can be built and operated by private companies in compliance with market rules, the strictest safety requirements, and appropriate provisions for nonproliferation.

Nuclear Energy Worldwide

Today, more than 440 nuclear reactors with a total installed capacity of 370 gigawatts (GW) provide 16 percent of the world's electricity in thirty-one countries to a population representing two-thirds of the planet. In the United States, 103 power reactors provide 20 percent of the country's electricity, and in Japan 52 reactors provide 34 percent. Half of the EU member states use nuclear energy, which, at 35 percent, is the main source of electricity generation in the EU.[20]

Almost thirty reactors, equal in capacity to 6 percent of current installed capacity, are currently being built in twelve countries including China, Korea, Japan, and Russia. Furthermore, there are confirmed plans to build a further thirty-five reactors, equal in capacity to 10 percent of current electricity production. Most of them will be built in Asia, an area characterized by high-growth economies and rapidly increasing electricity requirements.

China, which currently operates nine reactors, plans to quadruple its nuclear capacity by 2020 compared with what it is already using or building by connecting a further twenty to thirty reactors to the grid. Recently, the Chinese and U.S. governments reportedly signed a memorandum of understanding concerning four advanced third-generation nuclear reactors, including a technology transfer package. In India, the government is planning a nuclear program similar to that of China.

20 Of the thirty-one countries that use nuclear energy worldwide, seventeen are dependent on it for at least one-quarter of their electricity requirements. Within the EU, three-quarters of the electricity generated in France and Lithuania is nuclear, with one-third nuclear-based in Slovakia, Sweden, and Slovenia and two-thirds in Belgium. The figure for Germany and Finland is just over one-quarter.

Last year the Indian government gave its agreement in principle for the construction of eight additional reactors. Their construction could be granted to foreign companies but would require the removal of certain import restrictions, which could result from an international safeguards agreement in the context of nonproliferation.

In the United States, under the Energy Policy Act of 2005,[21] provision has been made for new reactors, some of which are currently being approved or ordered. Up to fifteen new nuclear power plants with a total capacity of 23 GW could be in operation by 2015.

Within the EU, several states have played a fundamental role over the years in the technological developments that have ensured the industrial and economic maturity of nuclear energy. Today, however, Europe is having difficulty establishing a collective energy policy and is dealing with these problems in a disorganized fashion. Three member states—Sweden, Germany, and Belgium—have approved a policy to gradually phase out nuclear energy although to date very few reactors have actually been shut down. At the same time, the first European pressurized reactor (EPR)—an advanced new-generation reactor featuring an evolutionary design integrating worldwide experience in operating light-water reactors, the results of R&D, an increased level of safety, better performance, and higher output—is currently under construction in Finland. The construction of a second is starting in France under the leadership of the French utility EDF and a possible partnership with other European utilities that could benefit from a share of the energy generated while participating in the construction of the reactor. In the United Kingdom nuclear energy is back in the headlines with the government's *Energy Review,* which provides for the private sector proposing, developing, building, and operating a new generation of nuclear reactors to replace current plants and thus meeting the country's new electricity requirements and dwindling domestic natural gas resources.[22]

---

21 The act guarantees loans covering up to 80 percent of new-build construction costs, a twenty-year extension of the limited insurance mechanisms financed by the operators themselves, and a tax credit (1.8 cents per kilowatthour [kWh] for the first six gigawatthours [GWh] generated by the new reactors during the first eight years of operation, capped annually at $125 million).

22 *The Energy Challenge: Energy Review Report 2006* (London: Department of Trade and Industry, July 2006), 17.

Table 1. Cost Structure for Electricity Generation Technologies, percent

Type of cost structure	Nuclear	Gas[1]	Coal	Wind
Investment	50–60	15–20	40–50	80–85
Operation and maintenance	30–35	5–10	15–25	10–15
Fuel	15–20	70–80	35–40	0

Source: Estimated by Areva on the basis of various international studies.

1 Combined cycle gas turbines.

In a nutshell, beyond the interest in nuclear power expressed by numerous countries, resorting to this form of energy as a means of providing base-load electricity is an option that has been selected or confirmed by more than thirty-five other countries whose demographic, economic, and geopolitical weight is considerable. They represent advanced societies where industry and technology are highly developed. They account for almost two-thirds of the world's population. Some of these countries are involved in developing more advanced, safer, and high-performance technologies for a more efficient use of resources, in particular uranium.

Competitiveness of Nuclear Energy

Despite high capital costs, nuclear energy is a competitive solution for supplying base-load electrical power at a low marginal cost and with extremely high availability. Furthermore, because changes in fuel costs have relatively little impact, nuclear energy can play a major role in generating electricity while offering very good cost predictability.

Numerous studies have been carried out worldwide in recent years looking at the cost of nuclear and other sources of energy. These costs include construction costs, operation and maintenance costs, together with decommissioning and fuel costs, and they can be broken down as illustrated in table 1.

Investment and construction. Capital costs are the major component in the cost of nuclear electricity. Investment costs encompass overhead costs, which include preliminary design, engineering, procurement, construction, and owner's costs (site preparation and regulation), operator training, any delivery delays and price variations as well as interest during construction, which represents financial costs resulting from construction lead times and that can be up to 20–30 percent of the overall investment. Investment costs can represent up to 60 percent of the overall cost of nuclear energy. Recent international studies confirm Organization for Economic Cooperation and Development (OECD) estimates that set the cost of building an EPR-type reactor at between 1,300 and 1,800 euros per kilowatt. We believe that nuclear energy is competitive within this margin, which can potentially be lowered by the expected series effect.

Fuel. Fuel costs, a component in the cost of generating electricity, are significantly less for nuclear energy (4.4 euros per megawatthour) than for coal (between 14.7 and 22.1 euros per megawatthour) or gas (26.5 to 32.4 euros per megawatthour). Furthermore, the cost of uranium itself is only a fraction—about 25 percent—of the cost of nuclear fuel, and fluctuations in the cost of raw materials have little impact on the overall cost of producing nuclear electricity. A doubling in the price of fuel would increase the marginal cost of electricity generated by a gas-powered plant by 70–80 percent while the cost of nuclear electricity would increase only by 5 percent.

Note that the geographical diversification of uranium deposits increases the guarantee of its availability. The IEA considers that "uranium resources are not expected to constrain the development of new nuclear capacity and that identified uranium resources are sufficient for several decades of operation at current usage rates." In the IEA scenarios, "all demand to 2030 can be met from reasonably assured resources at a production cost below 80 $/kg. Beyond 2030, the additional demand can still be met, on the basis of current estimates of total uranium resources, including reasonably assured, inferred and undiscovered resources."[23]

Operation and maintenance. Operation and maintenance costs can vary depending on the companies operating the reactors and on national requirements (including cost of labor, insurance, maintenance investment, and operator strategies). For the most part, these costs are not closely linked to the size of the reactors or the quantity of electricity generated, so operators try to ensure high availability of the installations. This can happen only if operations (fuel replacement, inspection, maintenance) are forecast and planned. As its marginal cost is

23 *World Energy Outlook 2006.*

Table 2. Operating Costs (Base-Load Electricity), 2001, euros per megawatthour

Country	Nuclear	Gas[1]	Pulverized coal
France	7.1	5.1	8.7
Finland	7.2	3.5	7.4
United Kingdom	7.9	4.7	4.7
OECD	6.0–9.1	4.6–5.1	6.6–8.7

Sources: For France: "Coûts de référence de la production électrique," DGEMP-DIDEME, Paris, December 2003; for Finland: R. Tarjanne and S. Luostarinen, "Economics of Nuclear Power in Finland," Lappeenranta University of Technology, Lappeenranta, Finland, 2001; for the United Kingdom: "Powering the Nation: A Review of the Costs of Generating Electricity," PB Power, Newcastle upon Tyne, March 2006; and for the OECD: *Projected Costs of Generating Electricity, 2005 Update* (Paris: OECD Nuclear Energy Agency, International Energy Agency, 2005).

1 Combined cycle gas turbines.

low, electronuclear electricity is at the top of the merit order, and the operator aims for full power as much as possible: availability in excess of 90 percent is the norm in reactors generating base-load electricity. Table 2 compares the operating costs of three electricity generation technologies; the comparison is based on four recent studies.

Decommissioning. As long as a long-term framework exists, as in many developed countries, facility dismantling is mainly a question of public confidence. As provisions are made during the service life of the plant—sixty years for the EPR, for example—the cost of decommissioning can be correctly forecast and managed, even if this cost can vary greatly from one country to another depending on the different public policies and reactor design. Decommissioning policies of various countries do not fundamentally alter the economics of nuclear power.

The cost of decommissioning existing reactors is about 240–900 euros per kilowatt.[24] For a new generation EPR-type reactor, the French utility EDF has estimated the total cost of dismantling at 420 million euros.[25] A special fund for raising this sum can easily be put in place

24 EDF recently estimated that the detailed cost of dismantling its four 900 MW pressurized water reactors on the Dampierre site (decommissioning, engineering, monitoring, maintenance, site security, waste treatment, and removal) would correspond to 15 percent of the total investment in real terms.

25 This number is based on technical data from Areva, Paris.

during the sixty years the reactor is in operation, with an annuity of about 2.8 million euros (which corresponds to around 0.14 euros per megawatthour). It is likely that these costs will drop later as experience is acquired during future decommissioning operations.

Competitiveness. Overall, compared with coal and natural gas, nuclear energy is competitive for generating base-load electricity. Four recent studies[26] confirmed this with regard to the cost of nuclear electricity compared with the cost of combined cycle gas turbines and the cost of coal combustion technologies even before taking the cost of carbon into account. When the cost of carbon is added, nuclear energy—even when taking into account decommissioning, used-fuel treatment, and waste management[27]—is clearly the most competitive solution.

Figure 1 shows the results of these national comparative studies for Belgium, Finland, France, and the UK. For each of the three technologies (nuclear, gas, and coal) the cost of electricity (euros per megawatthour) is broken down into the investors' share, operation and maintenance, fuel, taxes, and R&D, excluding any allowance for the cost of carbon. Since these reports were published, the cost of fossil fuel has increased significantly. Comparative costs of generated electricity, once corrected to take this into account, further increase the relative advantage of nuclear energy. As for gas, the price of which has doubled over the past four years, the impact on electricity is an addi-

---

26 See J. P. Pauwels and J.-M. Streydio, eds., *Report of the AMPERE (Analysis of the Means of Production of Electricity and the Restructuring of the Electricity Sector) Commission to the State Secretary for Energy and Sustainable Development* (Bruxelles: Commission pour l'Analyse des Modes de Production de l'Electricité et le Redéploiement des Energies, October 2000), http://mineco.fgov.be/energy/ampere_commission/home_fr.htm; R. Tarjanne and S. Luostarinen, "Economics of Nuclear Power in Finland," Lappeenranta University of Technology, Lappeenranta, Finland, 2001, www3.inspi.ufl.edu/ICAPP/program/abstracts/1184.pdf, a study conducted in Finland with a view to building the 5th nuclear reactor (2001); "Coûts de référence de la production électrique," DGEMP-DIDEME, Paris, December 2003, www.industrie.gouv.fr/energie/electric/se_ele_a10.htm, which takes the EPR as a reference; and *The Cost of Generating Electricity* (London: Royal Academy of Engineering, March 2004), www.raeng.org.uk/news/publications/list/reports/Cost_of_Generating_Electricity.pdf.

27 See the section of this paper about used fuel and waste management.

Figure 1. Comparative Electricity Costs for Belgium, Finland, France, and the United Kingdom

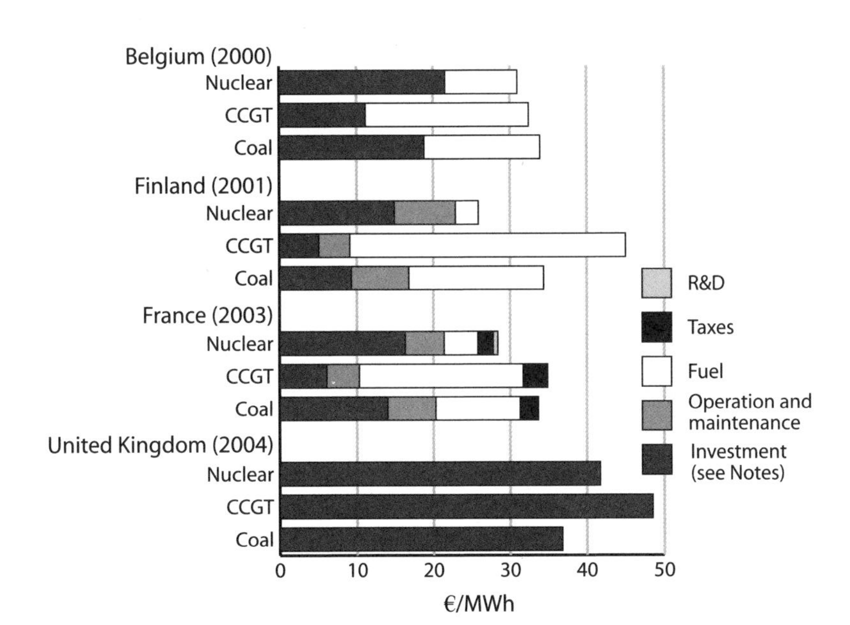

Sources: J. P. Pauwels and J.-M. Streydio, eds., *Report of the AMPERE (Analysis of the Means of Production of Electricity and the Restructuring of the Electricity Sector) Commission to the State Secretary for Energy and Sustainable Development* (Bruxelles: Commission pour l'Analyse des Modes de Production de l'Electricité et le Redéploiement des Energies, October 2000), R. Tarjanne and S. Luostarinen, "Economics of Nuclear Power in Finland," Lappeenranta University of Technology, Lappeenranta, Finland, 2001, "Coûts de référence de la production électrique," DGEMP-DIDEME, Paris, December 2003; *The Cost of Generating Electricity* (London: Royal Academy of Engineering, March 2004); and "Powering the Nation: A Review of the Costs of Generating Electricity," PB Power, Newcastle upon Tyne, March 2006.

Notes: For Belgium, operation and maintenance costs are combined with investment cost. For the United Kingdom, all costs were combined and are designated "investment" here. CCGT = combined cycle gas turbines.

tional 19 euros per megawatthour. For coal, the price of which has increased by 60 percent in western Europe in three years, the impact on the increase in electricity is 7 euros per megawatthour. Although the cost of uranium also increased, the effect on electricity has not exceeded 1 euro per megawatthour.[28]

Thus, the capital cost is a major factor in the cost price of electronuclear production. Recent work carried out by the OECD showed that for an average capital cost of 5 percent, the price of nuclear electricity fell within a range of 20–40 euros per megawatthour, whereas taking an assumption of 10 percent, these values could be between 25 and 60 euros per megawatthour.

Even on the basis of a rate of 10 percent, however, nuclear electricity remains competitive in numerous Trilateral countries, as illustrated by the results of national studies compiled by the OECD and shown in figure 2 for three sources of primary energy: nuclear, coal, and gas.

How investors perceive the risk also plays an important role in whether they decide to invest. Some studies require that nuclear projects have a greater return on investment and apply an additional risk factor of 3–4 percent to both the interest rate and the capital cost. We believe that there is no need to base the competitiveness of nuclear power on factors other than those used for gas and coal, especially as, objectively speaking, nuclear energy presents fewer risks in terms of the marginal cost of electricity, carbon costs, security of supply, and the weighting of the cost of fuel.

To conclude, by supplying electrical energy and not producing major amounts of CO_2 at a foreseeable cost and by offering protection against the cost of fossil fuels, nuclear energy offers additional advantages, which leads us to believe that it will continue to play a major role in the energy mix, notably in the recently deregulated markets.

Used Nuclear Fuel and Waste Management

Countries using nuclear energy have adopted different strategies regarding the management of used nuclear fuel. Some—especially France

28 These values are calculated taking into consideration the following hypotheses: For gas, the reference is a combined cycle plant with an output of 60 percent, the price of gas having risen from $3.00 to $6.00 per million BTUs. For coal, with an output of 42 percent, the fuel price increases from $35 to $55 per ton. The figures for nuclear energy concern an EPR, the price of uranium having increased from $20 to $40 per pound of U_3O_8.

Figure 2. Electricity Prices Compared in Selected Countries

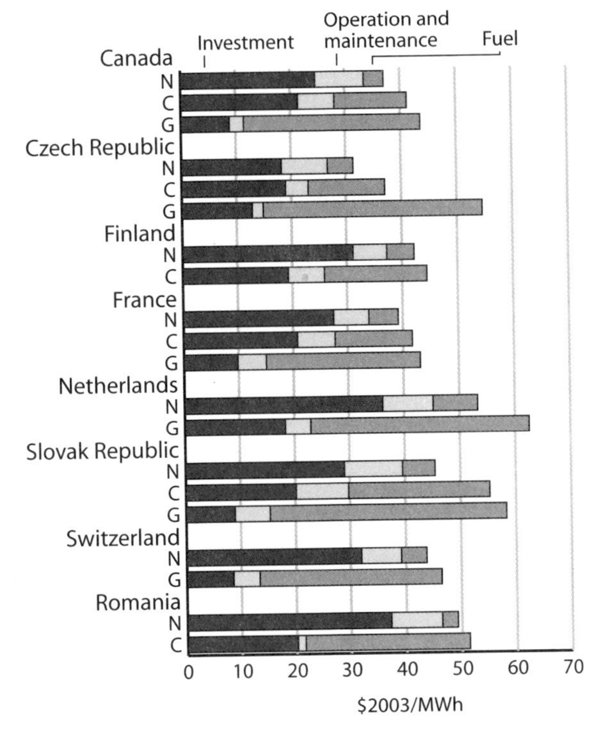

Source: *Projected Costs of Generating Electricity, 2005 Update* (Paris, OECD, Nuclear Energy Agency, International Energy Agency, 2005).

Notes: Selected countries are where mean levelized cost of nuclear in base load is the cheapest option (10 percent discounting). N = nuclear; C = coal; G = gas.

and Japan, but China, Russia, and India as well—pursue a recycling strategy in which used nuclear fuel is treated and then reused as a component of new reactor fuel. Used fuel can indeed be considered as a resource because it contains materials—uranium and plutonium—that can be recovered through treatment and recycling.[29] Other countries, notably the United States, pursue a once-through strategy in which

29 Thus 96 percent (95 percent uranium and 1 percent plutonium) of the material contained in a used fuel assembly can be recycled into fuel (for example, mixed oxide [MOX]) that can replace standard uranium fuel and save up to 25 percent of uranium resources.

untreated fuel is stored, later to be emplaced in a permanent geological repository.

There are four commercial operational treatment plants in the world; currently there are no operational repositories for commercial used fuel. To date, 20 percent of fuel already used in light-water reactors has been treated commercially; recycling in the form of mixed uranium and plutonium oxide (MOX) fuel has been organized on an industrial scale for twenty years.[30] In contrast, the conditioning and direct disposal of used fuel is still in the development stage in several countries,[31] with the estimated cost continuing to increase significantly. Residual waste from fuel packaged during the treatment and recycling process is vitrified in a matrix that is stable, homogeneous, and durable, no longer contains any fissile material subjected to the International Atomic Energy Agency (IAEA) safeguards system, and whose long-term behavior (several thousand years) is predictable, as opposed to fuel assemblies that are designed to be used only for a few years. Vitrification is now an industrially proven process for which industrial specifications and restrictions in terms of safety have been laid down by the national authorities in several Trilateral countries that possess this technology. On the basis of the principle, which is the law in France,[32] that all end-of-cycle waste is returned to the country of origin of the treated fuel, the authorities in each country involved have also drawn up and implemented broadly similar provisions relating to transport, acceptance, storage, and disposal of this "glass."

We consider that, viewed from an environmental and sanitary standpoint, treating and recycling used fuel has decisive advantages over disposal without treatment. Although reprocessing does not elimi-

---

30 Areva's industrial facilities have treated over 20,000 metric tons of used fuel over the past twenty years and have fabricated 2,000 MOX assemblies in the past decade.

31 In particular, the United States has pursued for the past twenty years the development of a geological repository solution for used fuel disposal at Yucca Mountain, Nevada, to handle all legacy fuel (estimated at 54,000 metric tons in 2005).

32 Under French national law, the return of all "foreign" nuclear waste, such as that resulting from the treatment on French soil of irradiated fuel entrusted by international customers, is a legal obligation (see act no. 2006-739 of June 28, 2006, relating to the sustainable management of radioactive materials and waste, article 8).

nate the need for managing final waste—the solution includes long-term interim storage and deep geological disposal—it reduces by 80 percent the volume of waste to be disposed of in a deep geological repository and by a comparable factor the radiological toxicity in the long term.[33] From an economic point of view, both strategies—the once-through cycle and recycling—are comparable. In addition, the cost of the back end of the fuel cycle represents only a small fraction of the total cost of the electricity generated—from 2 to 6 percent of the cost of the kilowatthour, according to various studies.[34]

Several factors have recently led to questions in the United States about the appropriateness of the once-through cycle as an exclusive used-fuel management strategy.[35] The economics of the back end of the nuclear fuel cycle and of developing a recycling strategy in the United States was recently reviewed by an independent study.[36] Recycling, as part of a portfolio strategy in which an integrated recycling plant complements the planned repository, was found economically competitive for solving the long-term used-fuel management requirements of the U.S. nuclear power market. Recycling shows economics comparable with an exclusive once-through strategy, especially considering uncertainties that surround many of the variables used in the assess-

---

33 This reduction in toxicity results from a scenario in which the used fuel from light-water reactors (with a burnup of 45 to 60 Gigawattday per ton is treated three years after being unloaded from the reactor.

34 These conclusions are taken from the following studies: *The Economic Future of Nuclear Power* (Chicago: University of Chicago, August 2004), www.anl.gov/Special_Reports/NuclEconSumAug04.pdf; *Reference Costs for Power Generation* (Paris: DGEMP/DIGEC, Ministère de l'Industrie, April 1997), www.industrie.gouv.fr/energie/electric/cdr-anglais.pdf; Paris; *The Economics of the Nuclear Fuel Cycle* (Paris: OECD, 1994), www.nea.fr/html/ndd/reports/efc/EFC-complete.pdf.

35 The changes concern inter alia the cost increase of the strategy, the likely long-term increase in nuclear generation, the expected need for additional repository capacity, and the economics of alternative used fuel management solutions, notably on the grounds of experience and industrial know-how built up during the past forty years.

36 *Economic Assessment of Used Nuclear Fuel Management in the United States* (Paris: Boston Consulting Group, July 2006), www.bcg.com/publications/files/2116202EconomicAssessmentReport24Jul0SR.pdf.

ment, such as uranium price and final repository costs.[37] Both factors have recently evolved in a way that makes the results of the study even more compelling for recycling. Moreover, such a strategy would offer the nuclear power sector protection against a potential rise in uranium prices by providing recycled fuel.[38]

We consider that opting to close the nuclear fuel cycle by treating and recycling used fuel is justified from both the economic and environmental points of view, and it helps preserve natural resources. This type of management ensures safe management of radioactive waste and reduces the constraints on future generations. It contributes to the sustainability and public acceptance of nuclear energy in the world.

Proliferation Issues

Nuclear nonproliferation policy aims to prevent the spread of weapons-grade materials (through diversion or undeclared production) and associated technologies. Because motivations for proliferation are political or geopolitical, states by definition hold the primary responsibility for enforcing nonproliferation. The international nonproliferation and, in particular, safeguards regime (for example, independent material accounting, containment and surveillance, and monitoring and environmental sampling) has been quite effective in dissuading potential proliferators. In addition, the need for an extended framework for physical protection has been recognized and taken into account under the Convention on the Physical Protection of Nuclear Materials. While the responsibility for implementing physical protection measures resides at the national level, the majority of countries, fully aware of the risks, have reassessed the threats and adapted the measures taken in connection with the latest international events.

Industry's key responsibility is to economically and efficiently ensure the supply of reactor and fuel-cycle materials and services in strict compliance with the international nonproliferation framework. With nuclear power poised to expand worldwide, the industry is in a unique

---

37 Note that some other issues need to be addressed, in particular the broad-based acceptance of recycled fuel by the nuclear industry as well as a positive legislative, policy, and financial environment for recycling.

38 MOX and recycled uranium oxide (UOX) fuel resulting from this strategy is estimated to be potentially able to satisfy 20–25 percent of U.S. fuel requirements.

position to contribute to the implementation of pragmatic, concrete solutions that facilitate expansion while limiting the proliferation risk.

Nonproliferation policy must be implemented and assessed in a holistic system-wide manner, employing a combination of technical (intrinsic) and institutional or organizational (extrinsic) measures, and addressing the specific context and overall system in which a given technology is used.

Recent developments in nonproliferation policy have focused on maximizing the intrinsic barriers of an isolated technology. While progress in this area is important and desirable, it is unrealistic and dangerous to believe that an intrinsically proliferation-proof technology can be developed and safely deployed in an isolated manner. Like safety, nonproliferation measures are not absolutes but instead are relative notions that evolve over time:

> . . . simply placing [used] nuclear fuel into a geologic repository does not "solve" the non-proliferation problem. The radiation barrier surrounding the [used] nuclear fuel continually decays away [leading to] what some people refer to as a "plutonium mine" if left in place long enough. The intrinsic proliferation resistance of the once-through cycle clearly decreases with time.[39]

The inherently political origins of proliferation mean that barriers extrinsic to technology will always play a fundamental role, and they deserve commensurate attention. Nevertheless, the closed fuel cycle or recycling option demonstrates several advantages. Waste conditioned in current industrial treatment facilities does not contain fissile materials subjected to the IAEA safeguards system. Furthermore, following its irradiation in the reactor, the quantity of plutonium that has been recycled in the form of MOX fuel is not only reduced by one-third, but the isotopic composition of what is left in the used fuel is significantly degraded so that the ability to use it in a weapon after it has been unloaded from the reactor is further reduced. Furthermore, improved

39 Alan E. Waltar and Ronald P. Omberg, compilers, "An Evaluation of the Proliferation Resistant Characteristics of Light Water Reactor Fuel with the Potential for Recycle in the United States," Pacific Northwest National Laboratory, Richland, Wash., 2004, 21, www.radiochemistry.org/documents.html.

technology is available for future investment that could contribute to a widening of access to nuclear energy for newcomers, without extending the use of sensitive technology. One example of this is the COEX process,[40] which avoids the separation of plutonium. This technology thus would make it even more difficult to use these materials for proliferation purposes.

Nuclear weapons have never been procured by diverting fissile materials from declared commercial treatment and recycling facilities. However, this does not in any way affect requirements in the industrial chain designed to ensure protection from proliferation. For instance, the treatment-recycling chain can be integrated into a broader nonproliferation scheme, in which some facilities under multinational control carry out treatment for several countries and the recycled fuel is made available to those countries under international guarantees. This would avoid the unnecessary dissemination of sensitive technologies and the continuous increase in plutonium stocks separated or contained in used fuel.

A multilateral approach to the fuel cycle was put forward in 2004 at the instigation of the IAEA. It aims to support the development of nuclear energy around the world through observance of the nonproliferation requirements relating to activities involving the civilian nuclear fuel cycle. In this way, it will establish guarantees for the supply of services (enrichment, processing, and recycling) for those countries that would not have developed domestic capacity through existing suppliers, and possibly in the long term also through the involvement of multilateral nuclear centers.

The nuclear industry recognizes and accepts its responsibilities in support of government directives.[41] Indeed, it is essential to nuclear commerce that customers are in full compliance with the safeguards regimes, and a more clear-cut penalty system for noncompliance should be agreed internationally at the intergovernmental level. The nuclear industry should, however, consider that market mechanisms have demonstrated their effectiveness in ensuring the continuity of these ser-

---

40 COEX is a process developed by Areva in association with the French Atomic Energy Commission.

41 *Ensuring Security of Supply in the International Nuclear Fuel Cycle* (London: World Nuclear Association, April 2006), www.world-nuclear.org/reference/pdf/security.pdf.

vices in favor of energy production. Any more restrictive new approach should, in the future, be compatible with the operation of a competitive worldwide market. In other words, the triggering of emergency measures, backup procedures, or guaranteed supply arrangements should occur only in the event of a political disruption of the normal market for a reason other that a nonproliferation issue.

The United States proposed a strategy called Global Nuclear Energy Partnership (GNEP), a vision associated with a long-term R&D program, in order to enable the United States and some restricted partners to recycle used nuclear fuel, develop advanced reactors to eliminate high-level waste (actinides) separated by the treatment process, and make nuclear fuel cycle services available to other countries with a view to avoiding dissemination of sensitive fuel-cycle facilities and thus reinforcing the nonproliferation regime.

This approach, which assumes the closure of the cycle, represents a logical and welcome development for the United States. Following a twenty-year hiatus in the development of its nuclear power program, the United States has now begun to restart it at home and promote its development elsewhere. Naturally, this has aroused debate on the other side of the Atlantic about its cost, its feasibility, the possible consequences in terms of proliferation as well as its influence on the future of nuclear energy throughout the world. All things considered, it is natural that a political development of this importance should provoke conflicting points of view, but it seems at least the process is under way, which should not be surprising considering that it is no more nor no less than a question of improving still further the conditions for a safe and stable supply for the world's nuclear power stations.

The concept of cradle-to-grave fuel services included in GNEP does however raise the question of the return of irradiated fuel to its country of origin and the fate of the related final waste. Where one (or more) regional reception center—almost certainly under public sector ownership and control—is to be set up for this purpose in certain services such as providing states, it would be essential for these to be accessible to others under nondiscriminatory conditions if they are not to lead to major distortions to competition in a market that must remain competitive in order to remain safe. Besides its still uncertain future on this point, this approach also fails to explain how the authorized partners might be approved to participate in the program to supply fuel-cycle services or the associated methods of control. Finally, any such approach must remain flexible: Depending on the development of its economy and nuclear program over time, a state that has initially made use of the guarantees of a supplier state must be able, when the time is right, to decide to supply itself autonomously if it considers that the scale of its activities will enable it to cover the cost and if it fully satisfies the international requirements relating to nonproliferation.

The Nuclear Way

Nuclear energy—currently the main source of electricity in the European Union—is a proven, competitive, and safe technology for the production of the base-load electricity required for the permanent needs of industry, the public transport sector, and many domestic applications. It also contributes to the satisfaction of the substitution requirements that occasionally arise from most renewable energy sources. Furthermore, nuclear power does not produce greenhouse gases, and its production costs are broadly foreseeable. It is the only energy source to have included for many years, in most Trilateral countries, its external costs (insurance, dismantling, and treatment and packaging of waste). New reactors can be built and operated by private companies in compliance with the rules of the market and the strictest safety requirements. Furthermore, current international R&D work will lead to the production of new, even safer reactors that will make better use of the energy potential of uranium and that will result in a very significant reduction in the volume of waste and its residual toxicity.

In this field, Europe, in particular, has the industrial structure, skills, and outstanding expertise at a time when all around the world—and especially among Europe's business competitors—very substantial investment is being made in the production of electricity. Because Europe is not an island lost in the middle of a calm ocean, Europe cannot afford to turn its back on these enormous worldwide business opportunities and on a source of energy that has the overwhelming support of its competitors, which will increase their comparative advantage.

The high capital costs of nuclear energy must be balanced against the lower vulnerability of the costs of the electricity produced compared with other primary energy sources (such as gas or coal), especially if carbon costs are included.[42] Furthermore the geographical availability and relative abundance of uranium resources throughout the

42 See the section in this chapter on the climate change issue.

world add to the attractiveness of nuclear energy for base-load electricity production. The long-term future must also be ensured by making the choices demanded by the end of the cycle and the disposal of ultimate waste, guaranteeing that the requirements imposed on the safety of installations are fully satisfied and the risks of proliferation are managed.

We think that these objectives are within the reach of the governments of the Trilateral Commission countries and that the public could be persuaded to sign up for them. As far as nuclear energy is concerned, a balanced series of proposals for a European Union energy policy should, at a minimum,

- Stress the importance of the current contribution of nuclear energy in satisfying competitiveness, security of supply, and sustainable development objectives;
- Encourage the immediate use of this form of base-load electricity production, while respecting the principles of subsidiarity;
- Add an overall contribution by nuclear energy to a target for low-carbon energy sources as part of the EU's energy balance, as well as quantified domestic energy targets;
- Promote a stable regulatory framework for the licensing, construction, operation, and dismantling of nuclear installations;
- Consolidate European technological and industrial leadership, notably through active involvement in international R&D activities dedicated to fourth-generation reactors, financing the establishment in Europe of essential research tools, and the construction of innovative demonstration facilities;
- Pursue R&D relating to improving the efficiency of the fuel cycle, in particular as it affects the volume of ultimate waste and its radiotoxicity;
- Support the efforts of governments getting involved in discussions and work at international level, with a view to a multilateral approach to the cycle (enrichment and treatment) that contributes to the development of nuclear energy in compliance with better management of the risks of proliferation; and
- Ensure the transfer of essential knowledge through information, education, and training programs.

Conclusion

Improvements in our knowledge of the worldwide risks posed by climate change confirm the scale of the problem and the urgency of the measures that must be taken. This action must be at an international level, in the field of government regulation, in terms of technological development, and, more globally, in our relationship with energy. The cost of measures designed to temper the effects and adapt to them is within our range; in any event it will be a great deal less than the cost of doing nothing. All affected countries, whether contributing to emissions or suffering the consequences, or both, must now stand together. They must all be involved in reaching rapid agreement on fairly dividing their efforts toward reducing and limiting their emissions and supporting the development of reliable, non-carbon energy technologies—including nuclear energy—in favor of more sustainable development. The United States will have a vital role to play in this respect. The Trilateral Commission, by its very nature and history, could provide a timely and appropriate path to such a consensus.

3

Pacific Asia Energy Security Issues

Widhyawan Prawiraatmadja

The Pacific Asia region is far from being a cohesive and homogenous group of countries. The diversity is arguably more immense than any other region in the world, as these countries face a myriad of differences in terms of economics, politics, demography, race, culture, and geography. What most countries in the region have in common, however, is their heavy dependence on external energy sources. The situation is more pronounced in oil, and even oil-producing countries in the region rely on imported supplies in a relatively significant amount. As oil is a strategic commodity with higher prices and inherent price volatility, the region's concerns about energy security are more than warranted.

Given the diversity among Asian countries, it comes as no surprise that there is still no cohesive policy and strategy in approaching and ensuring energy security. Fortunately, most key countries in the region are aware that the existence of such a policy and strategy is in the best interest of the region as a whole. As such, efforts toward achieving this goal have been made through cooperative initiatives involving certain country groupings and associations, such as the Northeast Asia Energy Policy Forum and the ASEAN + 3 initiatives. It is well understood that energy security should not be the concern of a single country, but rather a collective effort among the countries in the region or even across regions.

Pacific Asia Energy Supply and Demand Imbalances

Since surpassing North America in energy consumption in 2002, Pacific Asia is now the largest energy-consuming region in the world (figure 1), and its consumption growth continues to be faster than any

Figure 1. Regional Energy Consumption, 2005

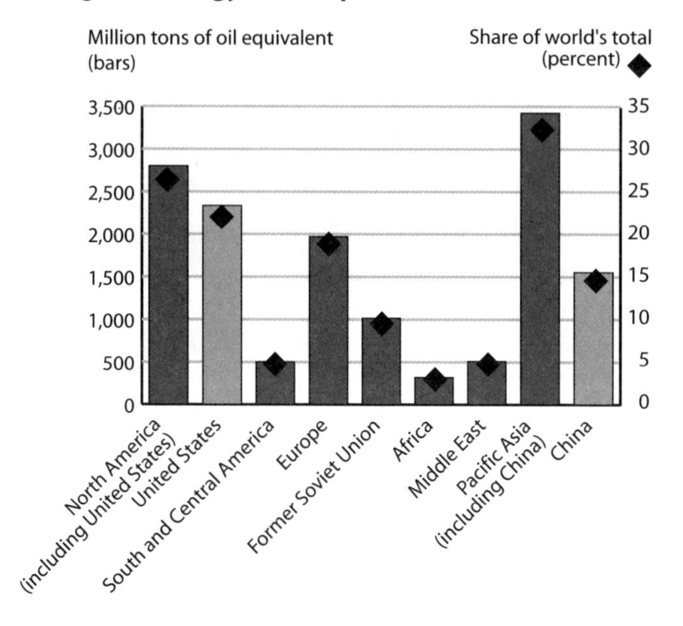

Source: *Statistical Review of World Energy 2006* (London: BP, June 2006).

other region (figure 2) despite the setback caused by the 1997–98 Asian financial and economic crisis and China's temporary reduction in coal use.[1] Along with economic development, the region's steady growth in energy consumption is expected to continue, especially in the Asian developing countries where energy consumption per capita is still very low (figure 3).

Unfortunately, most of the countries in the Pacific Asia region are not endowed with an adequate indigenous energy supply and, as a result, have to depend on imports. This situation is more pronounced in oil, as even the region's major oil producers such as China, Indonesia, Malaysia, India, Australia, and Vietnam are either net oil importers or will become importers in the near future. Further, as the oil re-

---

1 Unless otherwise specified, the data for primary energy in this paper are from *Statistical Review of World Energy 2006* (London: BP, June 2006). As such, we adopt the definition of primary energy as described in the review, that is, only commercial energy.

Figure 2. Relative Growth of Primary Energy Consumption, World Compared with Pacific Asia and China, 1990–2005

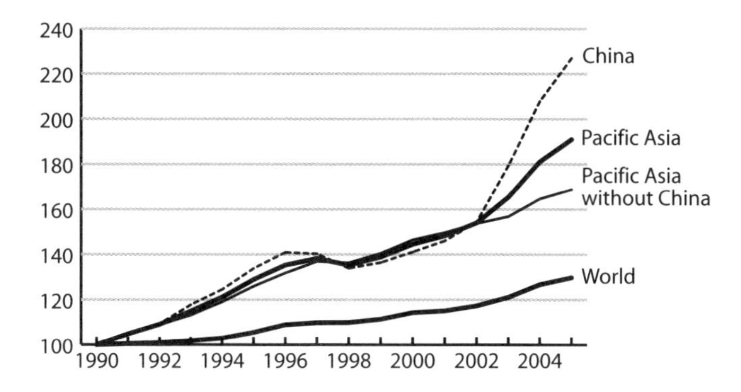

Source: *Statistical Review of World Energy 2006* (London: BP, June 2006).
Note: 1990 = 100.

Figure 3. Energy Consumption per 1,000 Residents, 2005

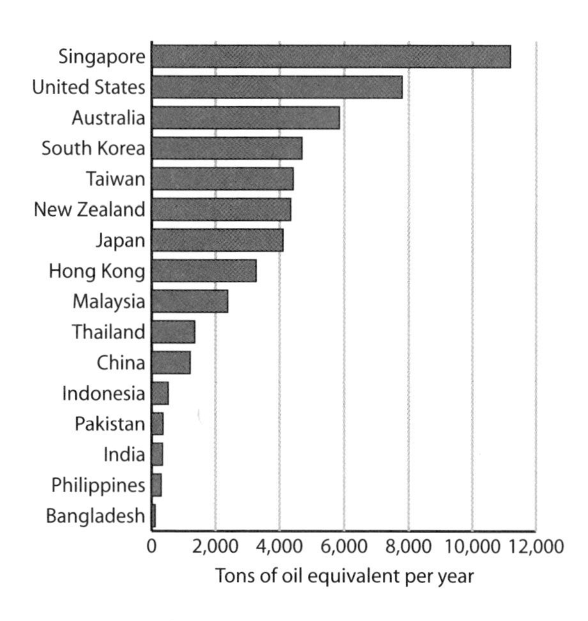

Source: *Statistical Review of World Energy 2006* (London: BP, June 2006).

Figure 4. Oil Production and Consumption, by Region, 2005

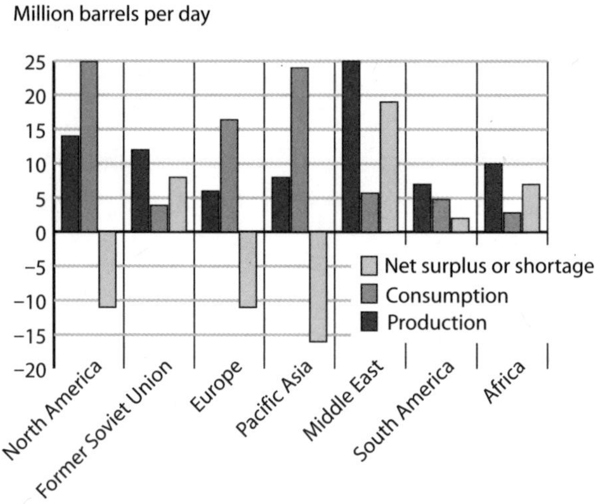

Source: *Statistical Review of World Energy 2006* (London: BP, June 2006).
Note: Negative numbers indicate shortages.

serves are limited, the region's oil production is not expected to dramatically increase, and there is a good chance that total production may indeed stabilize and eventually decline.

Figures 4, 5, and 6 show the Pacific Asia region's supply and demand imbalances in a global context for oil, natural gas, and coal. (The figures show whether a region is a net importer or exporter; as such, they do not show the intricacies of the intercountry trades within a region nor trades from a country in a deficit region to the other regions.) It can be seen from these figures that the region is almost in balance in coal, has a small shortage in natural gas, and is in a huge deficit in oil.

Pacific Asia Dependence on Oil Imports

Note that, given trends in oil production and consumption in the Pacific Asia region, the region's sizable dependence on Middle Eastern oil is a fact of life and is here to stay. This will continue in the foreseeable future despite all efforts in energy diversification away from oil, as well as oil supply diversification away from the Middle East. This is because of the Middle East's vast reserves and close proximity to the

Figure 5. Natural Gas Production and Consumption, by Region, 2005

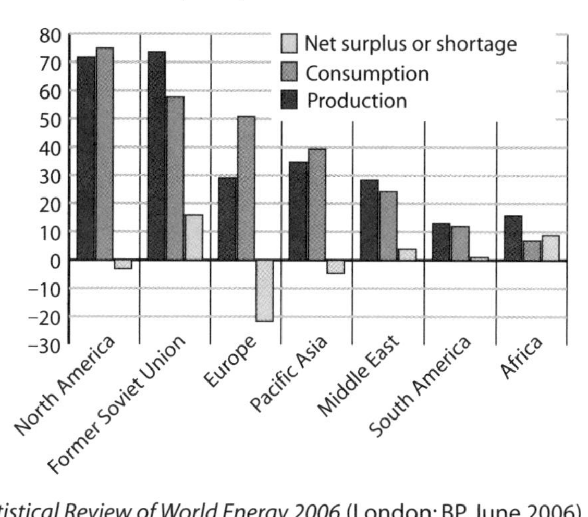

Source: *Statistical Review of World Energy 2006* (London: BP, June 2006).
Note: Negative numbers indicate shortages.

Figure 6. Coal Production and Consumption, by Region, 2005

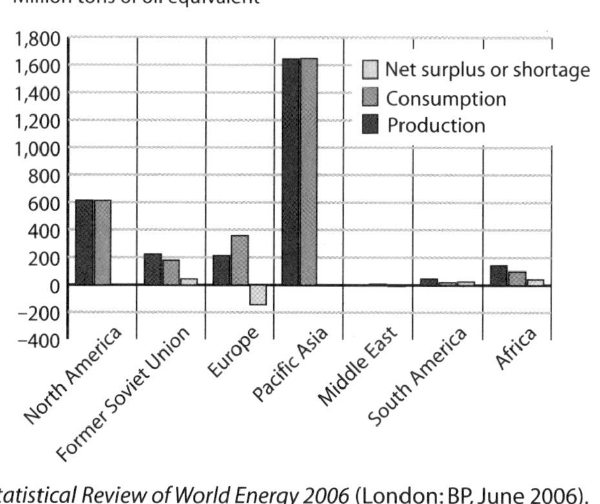

Source: *Statistical Review of World Energy 2006* (London: BP, June 2006).
Note: Negative numbers indicate shortages.

Table 1. Pacific Asia Crude Oil Imports, Including Share from Middle East, 2005–15, est., thousands of barrels per day

	2005	2006	2007	2008	2010	2012	2015
Crude runs	20,758	21,599	22,281	22,962	24,797	26,002	27,235
Crude direct use	398	432	429	425	408	394	375
Total crude demand	21,156	22,031	22,709	23,387	25,205	26,396	27,610
Asia-Pacific crude output	7,499	7,730	7,971	8,149	8,326	8,410	8,409
Own crude use	5,445	5,758	5,865	6,112	6,390	6,543	6,548
Crude imports	15,711	16,274	16,844	17,275	18,815	19,943	21,062
From within region	1,854	1,773	1,926	1,877	1,800	1,821	1,725
From Africa, Americas, Europe	2,415	2,415	2,525	2,760	3,150	3,250	3,500
From Middle East	11,214	12,086	12,393	12,637	13,865	14,872	15,837
Middle East share (percent)	73	74	74	73	74	75	75

Source: FACTS Global Energy.

Pacific Asia region, which has become and will continue to be the natural home for Mideast crude oil. The Middle East and the Pacific Asia region have become one regional oil market—the east-of-Suez market.

Table 1 shows the Pacific Asia region's growing crude oil imports. To complement Mideast crude oil, the Pacific Asia region has brought in supplies from other regions such as Africa, South America, and Europe (including Russia). This is mostly because of crude quality requirements in the Pacific Asia refineries. The growing needs for crude oil indicate that the Pacific Asia countries have increased and will continue to increase their refining capacity. Indeed, for energy security reasons, most countries prefer to refine the imported crude oil at home instead of relying too much on imported petroleum products. As such, over the years, net petroleum product imports have actually been quite stable, at around 2 million barrels per day.

Figure 7. Mix of Primary Energy in the Pacific Asia Region, 2005

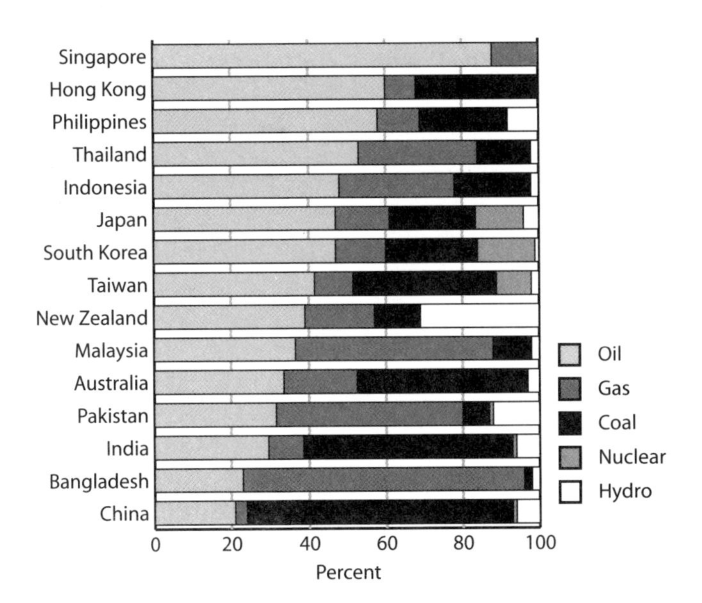

Source: *Statistical Review of World Energy 2006* (London: BP, June 2006).

Pacific Asia Energy Vulnerability

The Pacific Asia region's energy system is vulnerable in many aspects—in primary energy in general and in oil in particular. In the context of primary energy, the vulnerability includes the countries' reliance on certain energy sources, the pull of Asian resources to other net importing regions (such as to the United States), and the countries' competition in securing energy supplies. The vulnerabilities related to oil include the high reliance on Middle East oil, the volatility of (imported) oil prices, securing the ever-growing oil imports through congested sea lanes, the existence of market protection, subsidies and half-hearted deregulation in domestic oil markets, different levels of fuel-quality standards, uneven strategic oil stocks among countries, and the lack of a futures market. Some of these issues are elaborated below.

Countries in the Pacific Asia region possess a wide range of fuels in the primary energy mix. There are countries that arguably rely heavily on particular energy sources in their energy mix (figure 7), whether in the overall primary energy mix or in electricity generation.

In terms of supply, a few countries in the Pacific Asia region export energy resources to other countries, including those outside the region. As previously mentioned, oil exporters in the Pacific Asia region also rely on imports. With relatively higher reserves in coal and, to a lesser extent, natural gas, exports of these fuels to other countries in other regions is also apparent. The key to this is that exports to other regions mean that there will be a reduction in potential future supply to the Pacific Asia region itself. For example, growing shortages of natural gas in the United States will ultimately draw substantial supplies from Pacific Asia liquefied natural gas (LNG) exporters such as Indonesia and Australia.

Some Pacific Asia countries, which are experiencing steady growth in energy demand, are securing energy supplies while enhancing their energy supply security (such as in their efforts to diversify their suppliers). These countries essentially compete with other countries in the region in securing such supplies. This is apparent in the case of attracting imports of crude oil and natural gas from Russia and in supplying the Northeast Asian countries such as China, Japan, and South Korea. The competition also extends to other (potential) buyers in other regions, such as the United States and western European countries.

As to vulnerabilities specifically related to oil, the heavy dependence on oil imports, especially from the Middle East, brings consequences in potential supply disruptions because of never-ending conflicts and instability. As oil imports continue to grow, securing the ever-increasing oil supplies that require passage through crowded sea lanes is becoming more of a challenge. Congested sea lanes like the Malacca Strait are more prone than other areas to accidents, piracy, and even the possibility of terrorist attacks; hence the increase in the chance of a supply disruption.

In a final line of defense in the event of supply disruptions, some countries have established strategic oil reserves. The more affluent countries can afford to have strategic reserves and stockpiles, of course, whereas developing countries are more exposed. The government of Japan, for example, holds about ninety days worth of oil consumption in national strategic reserves managed by the Japan National Oil Corporation (JNOC). In addition, Japan's government also requires private companies to hold substantial crude and petroleum product stocks. South Korea maintains reserves equal to ninety days of consumption. Taiwan is also relatively well prepared for potential supply disruptions. It requires both the state-owned Chinese Petroleum Corporation (CPC) and the privately owned Formosa Petrochemical Corporation (FPC) to have at least sixty days worth of their sales in storage. Finally, as a regional refining center and oil trading hub, Singapore has large commercial stocks on hand at any given time. It also requires its three state power companies to each have ninety days of oil consumption in storage. Other countries aspire to have a certain level of strategic stocks and also increase them, but most have been prevented from doing so by current higher oil prices.

Last but not least, the Pacific Asia region badly needs a viable oil futures exchange (similar to Nymex in North America and IPE in Europe) that caters to the region as a whole instead of only a particular country (for example, the Tokyo Commodity Exchange [TOCOM] in Japan) to establish a real linkage to the global market. This would enhance oil market transparency and allow countries and companies to more effectively lessen the negative impacts from oil market disruptions and price volatility, if they desire. It is important to note that the oil futures market plays a critical role in the global oil market. Today, the number of paper contracts that are traded is seven to eight times larger than the volume of physical crude. The constant turnover of contracts leads to better price clarity for both buyers and sellers.

Areas for Cooperation and Collaboration

If close energy trade ties between Asia and the Middle East have been formed mainly out of commercial needs, then investment ties are partly pursued by governments in both regions to ensure energy security: solid market outlets for Middle East exporters and stable oil supply for Asian importers. This does not suggest that a government push is the only reason behind cross-regional investment activities; it is merely an indication that uninterrupted oil flows from the Middle East to Asia are very important to countries in both regions.

Over the years, there have been successful efforts in oil-related investments by Asian countries in the Middle East and vice versa. These investments have been characterized primarily by Asian countries' investments in upstream activities in the Middle East and Mideast countries' investments in downstream projects in Asia. This has further integrated the two regions in economic terms. As far as oil flows are concerned, Mideast oil producers have as much to lose financially from interruptions in the oil trade as do Asian buyers. Patterns of mutual investments have created equity partnerships that foster reliable oil flow and have thereby enhanced energy security within the Pacific Asia region. This cooperation can and should be extended to include storage capacity for Middle East crude oil in Asia.

Despite their fast-growing exports to Asia, Middle East countries do not have substantial crude storage facilities in place in the region. Middle East countries already have such facilities in place in Europe and the Caribbean. There is clearly a need for crude storage, in both Southeast Asia and East Asia. Storage facilities could store large volumes of Middle East crude for easy access to Asian markets. Such storage in Asia is essential for both commercial and security reasons. For such a project to succeed, key Middle East countries such as Saudi Arabia, Kuwait, the United Arab Emirates, and possibly Iran must participate together with Asia's private sector or state oil companies. (Because of Saudi Arabia's large exports and significant equity partnerships in Southeast and East Asian refineries, there is a particularly strong logic for Saudi involvement in a large storage facility in the region in the first instance.) Asian governments should pursue this option and encourage Middle Eastern major oil suppliers to collaborate by facilitating the construction of such storage facilities. This will go a long way toward reducing oil security concerns.

Indeed, the importance of additional stockpiles is imminent vis-à-vis the ever-increasing volumes of imported oil to Asia, especially in growing markets such as China, India, and other Southeast Asian countries. This is especially important considering the disruption that can happen in sea transportation lanes. Cooperation and coordination among Asian countries should be sought to make this materialize.

Internally, Asian governments should coordinate policies and encourage countries without strategic stockpiles to build such stockpiles. These stockpiles could be held offshore in jointly owned storage or within national borders. To the extent possible, Asian governments should be encouraged to move toward International Energy Agency (IEA)–style rules—in a step-by-step approach—for building strategic stockpiles and depleting them in the case of emergency. It is very important that stockpiling policies be used in a regional context and not just by individual countries. Although financing for building facilities and the costs of carrying stocks may be substantial, in the end the coordinated efforts may prove to be more cost-effective than individual countries' strategic reserves combined. Certain individual governments may need initial assistance, and getting such mechanisms in place may be worthwhile to the more affluent Asian countries such as Japan.

There are, of course, challenges to cooperation, coordination, and collaboration. In recent years, we witnessed an important (gradual) attitude change in the Pacific Asia region: Some countries in the Pacific Asia region aspire to have stronger commercial ties with Middle Eastern governments, mostly because of the inevitable energy links between the Middle East and Asia. This changing relationship has had an impact on the relationship of both regions with the Western powers, particularly the United States.

Establishing cooperation among countries within the Pacific Asia region has confronted several challenges, most notably internal politics and tensions between countries. These can pose unnecessary barriers to efficient energy procurement through market mechanisms in a cooperative manner. The continuing tensions between India and Pakistan have hindered more economically logical natural gas imports (compared with LNG) via pipeline from the Middle East to India via Pakistan. Similarly, internal politics in Bangladesh have inhibited the ability of India and Bangladesh to take advantage of cost-effective and mutually beneficial cross-border natural gas trades.

Concluding Remarks

Concerns of energy security in the Pacific Asia region originated from the region's substantial dependence on oil imports. Currently this has been exacerbated by higher prices. Energy security in Asia no longer pertains to energy importers only. In oil, even key crude oil producers, such as China and Indonesia, are major importers of petroleum products and crude oil. China, the largest crude oil producer in the region, has been a net oil importer for quite some time. Indeed, the emergence of two large countries in Asia—China and India—as major crude oil importers will significantly increase the Pacific Asia region's oil import dependence in the long run. As China and India have joined the ranks of the traditional large oil importers in the region, led by Japan, South Korea, and Taiwan, it becomes increasingly important for countries in the Pacific Asia region to cooperate in their strategies addressing issues of energy security.

The Pacific Asia region is much less dependent on natural gas than it is on oil. Despite its environmental benefits, the share of natural gas in the energy mix is still relatively low compared with other regions. This fact stems from the geography of Asia, which in many cases necessitates natural gas trading in the form of a higher cost for LNG and long-distance pipelines. Moreover, many consuming countries lack the requisite infrastructure to expand the share of gas in their overall energy mix. Nonetheless, efforts to increase the use of natural gas continue whenever long-term economic benefits warrant them. Indeed, expansion toward further use of natural gas in the primary energy mix is viewed as the desirable option in energy supply diversification. As far as coal is concerned, consumption in the region has no serious energy security implications. Despite higher pollution effects, coal consumption continues to be substantial because of its relatively low price and regional availability. This fuel will continue to play an important role, especially in the power sector.

Because the Middle East supplies most of Asia's oil, economic linkages between Asia and the Middle East are crucial to enhancing energy security for the Pacific Asia region. There have been successful efforts in oil-related investments by Asian countries in the Middle East and vice versa. A logical extension to current efforts would be collaboration in storage facilities in Asia that would serve Middle East suppliers' commercial purposes and enhance the Pacific Asia region's energy security at the same time.

As there are concerns about ocean transportation lanes, especially considering the ever-increasing volume of oil that needs to be imported, additional oil stockpiles have become an important issue because, if implemented, they would enhance oil supply security. This will actually benefit both producers and consumers because it would lead to solid market outlets for Middle East exporters and stable oil supplies for Asian importers. As far as oil flows are concerned, Middle East oil producers have as much to lose financially from interruptions of the oil trade as do Asian buyers.

Indeed, further cooperation and collaboration in the efforts to enhance energy security are needed, for all the countries within the region as well as other regions in a mutually beneficial manner. This is to be complemented by the harmonization of domestic energy policy in Pacific Asia countries with respect to deregulation and market reforms (particularly in the domestic oil market), fuel standards, and international trades.